A NEW VOYAGE: IMAGING THE NEXT ERA OF MARITIME TRANSPORT COMPANIES

By; Mustafa Nejem

TABLE OF CONTENT

ABOUT THE AUTHOR

The Author is a maritime visionary with a captain's heart and an island soul. Born into a lineage where, generation after generation, the love of the sea, the tradition of sailing, and the torch of leadership were revered and passed down with pride and glory in his island home. He's not just a mariner but also a wordsmith. Proficient in the navigation of language, he masterfully sails through the vast ocean of knowledge. With unwavering expertise and palpable passion, he guides readers toward prosperous shores, unveiling the secrets of maritime life and business success in a prose that is both concise and captivating. Drawing from his rich heritage and vast experience, this book is an evidence to his dedication to the maritime world.

PREFACE

This book was inspired by a deep admiration for the maritime industry and its vital role in global history, commerce, and culture. It explores the complex aspects of the marine operations, combining its extensive historical background with the progressive developments that are influencing its path. This book examines the symbiotic relationship between traditional maritime traditions and the necessity for innovation within the ever-evolving oceanic business, spanning from the experiences of ancient seafarers to the advancements of modern vessels.

The problems and opportunities within the maritime sector are reflective of the larger transformations occurring in the global landscape. The future of marine is significantly influenced by climate change, technology improvements, economic dynamics, and geopolitical tensions. With a recognition of the inherent interconnection, the primary objective of this book is to furnish a thorough comprehension, thereby bridging the divide between the marine industry and the broader global framework within which it functions.

For individuals with extensive experience in the marine industry, this book serves as a contemplative piece, offering valuable perspectives and serving as a navigational aid for future endeavors. For individuals who are unfamiliar with this field, it provides an entry point to a realm that, despite frequently being disregarded, holds significant importance in the intricate fabric of our interconnected global society.

This work is a tribute to the marine sector and the numerous individuals who navigate its journey. As you commence upon this literary expedition, it is our aspiration that you acquire not only knowledge but also a heightened viewpoint on an enterprise that consistently molds, and is molded by, our worldwide narrative.

Respected regards.

INTRODUCTION

From Age-Old Traditions to 2023

The history of the sea is as old as humanity. Many generations ago, our ancestors used to sail the seas and discover new lands. They traded, explored new lands, and connected disparate cultures. This eventually led to major shifts. For instance, when steam engines were introduced, ships could travel quicker and carry more cargo. The construction of the Suez and Panama Canal accelerated travel.

Over the past century, maritime traditions and technological advancements reshaped the industry. The evolution was evident not just in the size of the vessels but in their capabilities as well. Initially, efforts centered on making ships bigger to carry more cargo as international trade grew. This led to the building of huge ships, some of which were as long as four football fields put together.

However, by 2023, size wasn't the sole defining characteristic. Modern ships had become hubs of intelligence and efficiency. Onboard, cutting-edge technologies like Artificial Intelligence (AI) played pivotal roles. Advanced algorithms and systems assisted crews in navigation, optimized fuel consumption, and even predicted maintenance needs. It wasn't just about transporting goods anymore; it was about doing so with precision, safety, and sustainability in mind.

At the same time, ports went through distinct changes. Operations had to be streamlined because the number of objects transported was growing. Advanced management systems, supported by technologies like the Internet of Things (IoT) and automation, made this feasible. These systems could track cargo in real-time, automatically schedule shipments, and ensure the quickest, safest handling of goods. The collaboration between smart ships and smart ports promised a maritime industry more responsive and efficient than ever before.

Along with that, the rules for ships also changed. There were new standards for cleaner and safer seas. For instance, the companies had to think about the environment more and use cleaner fuel. Safety became a big deal, too, especially in places where piracy was a concern.

Fast forward to 2023, the sea world is now a combination of old and new structures. Once age-old tales and traditions are now taken over by new tech and challenges. And that's where our journey begins, looking at what's next for ships and the sea.

The Crucial Pivot: The Transformative Decade Ahead

The maritime industry is on the cusp of a major shift right now. These changes can be imagined as you are standing on the shore and seeing a huge wave coming. This huge wave symbolizes the massive changes that are going to occur in the maritime industry. The next 5-10 years in this industry are not going to be the same as usual. There are going to be great changes and big-time advancements.

Why? Well, there are a few reasons.

First, the state of the planet's ecosystem is an important concern. The maritime sector is committed to improving environmental sustainability alongside the rest of humanity. They are feeling the heat to become more environmentally conscious and reduce their carbon footprint.

Then there's technology. Just like smartphones have changed our lives, new technologies are changing the maritime world. We're talking smarter ships, better navigation, and even using robots for some tasks.

Lastly, the world itself is changing. Trade routes are constantly evolving, market dynamics are shifting, and the maritime industry needs to keep up.

This time, it is not just about making a few things right for a smooth sale. The industry needs to throw out the old, traditional rulebook and create a new one. It's about time to think big and plan for a future where the maritime industry flourishes and leads, not just merely survives.

Hence, the next decade in the maritime world is going to be a wild and exciting journey.

Purpose of this Book

When you set on a big sea adventure, you do not leave without a map. This book is like that map but for the future of ships and seas.

Now, this isn't just a book predicting what might happen. Nope, it's more like a wake-up call. We're telling everyone in the maritime world – from the pros at the helm to the people managing the show on land – "Hey, things are changing! Are you ready?"

The world of ships is about to see some pretty big waves of change. Companies like Maersk and MSC are the pioneers of surfing these waves, showing us the great possibilities. Analyzing their positive changes and backing them up with some authentic research, we will fill you in on what to expect in the years ahead.

However, this is not merely about knowing what's coming; it's about being ready for it. The goal here is more than to keep afloat; it is to sail ahead confidently. We want to ensure everyone has the map to tackle the challenges, grab the opportunities, and steer their ship in the right direction.

In short? This book is your loyal compass for the exciting journey ahead in the maritime world. Let's set sail into the future, prepared and ready for the storms!

Chapter 1

INTRODUCTION TO THE MARITIME TRANSPORT SECTOR

Salutations, and welcome onboard! We're setting off on a journey, not just across the high seas, but one that spans generations, technologies, and global boundaries. This journey is bigger than any one ship, port, or even nation. It's about how humanity has come to rely on maritime transport in more ways than most of us even realize.

Maritime is more than big ships carrying goods over vast oceans, although that's a big part of it. It's also about the unsung heroes who navigate these ships, the tech wizards who make them smarter, the environmentalists who strive to make them greener, and the entrepreneurs who make the deals that keep goods flowing across the globe. It's a complex symphony that involves a multitude of resources, all coordinated in a grand, global market.

The reason why it is important to learn about this growing industry is because maritime affects us in ways big and small. Whether you're sipping coffee that came from another continent, driving a car made from parts manufactured worldwide, or using a phone assembled in a far-off country, maritime transport made that possible. It's not just the backbone of the global economy; it is also the arteries, veins, and the pumping heart.

This industry is evolving every single day. It's dynamic, ever-changing, challenging, and exhilarating. The industry's constant change is what makes it fascinating.

Understanding the maritime world begins with this chapter. This book will be your guide to the constantly evolving maritime industry. It will be a thrilling voyage through history's tides and change, so sit back and hold onto your hats.

Evolution of Maritime Transport

Imagine a time before humans discovered they could levitate on water. Perhaps they began with a plank or a canoe. Those early adventurers were like the Silicon Valley startups of their day, breaking new territory and taking risks.

A short time later, sailing ships began to appear. Picture buried treasure, daring sea voyages, and pirates. Sails captured the wind and carried explorers to distant locations. But sails had their limitations; they were dependent on the wind, which does not always move in the desired direction.

Then came the era of the steam engine. This was a game-changing development. Now, ships could travel quickly without waiting for a windy day. For the first time in history, humanity had a dependable method for traversing oceans regardless of the weather. However, those steamships produced a lot of smoke and were threat to the sustainable environment.

Today, we have ships that are essentially floating structures. These vessels are so large that they have their own postal codes (not literally, but you get the idea). Moreover, these high-tech wonders aren't limited to burning smoky or toxic fuels any longer. Thanks to solar panels, wind turbines, and alternative fuels like hydrogen and ammonia, they're gradually becoming more environmentally friendly.

But let's be honest: it wasn't all sunshine. The maritime sector experienced its fair share of typhoons and icebergs. Wars broke out, impeding trade routes. Piracy still makes headlines, and economic fluctuations can upend everything.

Through it all, the industry continued to operate. When problems arose, individuals displayed creativity. They discovered new routes, designed smarter ships, and began focusing more on ocean cleanliness.

Hence, we stand in a period of transition. The maritime industry is no longer restricted to transporting goods from A to B. It has become more about doing it intelligently, efficiently, and effectively. And this is an excellent time to discover details about this ever-evolving industry.

The Role of Maritime in Global Trade

You do not need to be a sailor or ship proprietor to learn about the shipping and maritime industries. But if you enjoy having a variety of goods at your disposal, from the smartphone you can't exist without to the coffee that gets you going in the morning, then know that maritime is of great essence to you.

It is safe to say that ships are the unsung heroes that provide us with our daily requirements and wants. Here's a staggering fact: approximately 80% of all global commerce is transported by sea. In all likelihood, four out of every five of the goods you own were transported to you by ship in some capacity. If maritime transport went on strike, the situation would be comparable to a world without the internet or coffee. Grocery store shelves would dwindle, gas prices would skyrocket, and online purchasing would be nonexistent.

As stated previously, maritime is not just a hauling enterprise. It is a bridge between the countries. When cocoa beans are transported from Ghana to Belgium to be processed into chocolate, more than just products are being transported. It's a transmission of culture, of expertise, and of economic value. It involves creating employment in one location and meeting needs in another.

However, the maritime world is not merely a conveyor belt. It is also where important standards are established. Consider safety and environmental regulations. Ships must adhere to a number of regulations to ensure that they do not pollute the oceans or create hazardous conditions. And given that they operate in international waters, these regulations frequently serve as a model for larger global regulations.

And let's not overlook its influence on the economy. Ports are frequently the beating center of their cities, generating employment not only at sea but also on land. From dockworkers to customs officers to logistics planners, maritime transportation supports many occupations.

The importance of maritime shipping cannot be overstated; it is a crucial pillar upon which the global economy rests. If the maritime industry sneezes, the entire globe gets sick. It is the network that enables our contemporary world to exist as is.

The maritime industry is like the stage manager of a grand play, silently ensuring that everything functions smoothly in the background. And similar to any great production, when it is executed well, you barely perceive it. But when things go awry? Everyone feels the effect.

Spotlight on Market Leaders and Disruptors

Having acquired a foundational understanding of the marine industry, let us now focus on the Most Valuable Players (MVPs) within this sector. These companies are not merely participants in the industry; rather, they are actively reshaping its conventions.

Hypothetically speaking, these companies are the ones you want on your maritime team because they set the standards and lead the way. Consider them as the LeBron James of basketball or the Elon Musk of technology, but in the context of maritime pursuits.

The first enterprise in our hall of fame is Maersk.

Maersk is more than just a big name; it's like the quarterback everyone's watching. It doesn't just make moves; it sets the pace for the entire league to follow. You might say Maersk has been turning the wheels (or should we say propellers?) of the maritime world for quite some time now. If there were an Oscars for maritime companies, Maersk would be a frequent nominee, if not the winner, almost every year.

So why does Maersk get this star treatment? The company has been in the game for a long time, and it has an exceptional track record that most companies only idealize.

This successful corporation demonstrates a keen capacity to foster innovation, a steadfast dedication to sustainable practices, and an unparalleled global presence that distinguishes it from its competitors. And we are not referring to an organization that excels in only one area. Maersk demonstrates exceptional performance in various domains, encompassing shipping, logistics, sustainability, and technological innovation.

If you want to understand what leadership in maritime looks like Maersk is an exemplary case study. It's the company that both veterans and newcomers in the maritime industry look up to. It sets the trends, breaks the molds, and isn't afraid to steer the ship in a new direction.

Maersk: An Industry Leader

Maersk is a corporation that frequently writes the maritime playbook, not just follows it. When the company takes action, it garners attention and provokes interest. Let us further explore the distinguishing factors that differentiate Maersk from its competitors.

Brief History and Growth of Maersk

Consider Denmark during the year 1904. A.P. Moller undertakes a seemingly precarious endeavor by establishing a shipping firm with a single ship. In the present day, Maersk has evolved beyond being solely a shipping firm and has emerged as the preeminent entity in the maritime industry. To be accurate, it is the largest container transportation corporation in the world. This is not a game of chance; rather, it is growth achieved through careful planning and execution.

How did Maersk grow from a single ship to a prominent nautical entity? The company underwent not only expansion but also diversification. It is comparable to a gifted musician who is proficient in several musical genres and who also writes, produces, and possibly even directs music videos. Likewise, Maersk initially established dominance in the shipping industry. then expanded its operations to include the oil sector. However, the situation did not conclude at that point. Additionally, the corporation expanded its operations into the logistics sector, including the undertaking of port operations.

During this time, Maersk made some astute moves, such as forming strategic alliances and acquiring niche competitors. Maersk strategically incorporated these partnerships and acquisitions into its business plan as a means to maintain a competitive advantage, whether through the adoption of innovative technologies or the establishment of crucial shipping routes.

In short, Maersk epitomizes adaptability and innovation, seamlessly bridging traditional maritime practices with forward-thinking strategies. Their resilience and versatility set them apart as an industry leader in a rapidly evolving global sector.

Innovative Ventures and Future Plans of Maersk

Maersk is more than a shipping giant; it's a visionary. While many companies are still navigating the present, Maersk is already charting courses for the future. So, what's on the horizon?

Sustainability

Maersk is not just engaging in verbal discourse but also actively demonstrating its commitment through tangible actions. They've committed to zero carbon emissions by 2050. This isn't a small step; it's monumental. To reach this lofty goal, Maersk is investing in cleaner fuel options like biofuels and experimenting with technologies such as electric propulsion.

Technology

Maersk is jumping into the digital age with both feet. Think blockchain for more secure shipping and data analytics to optimize routes. They're not just streamlining their own operations; they're setting new industry standards.

But Maersk isn't going it alone. They're big on partnerships, whether it's joining forces with tech startups for fresh ideas or teaming up with local governments for more sustainable practices. In short, they're all about collaboration.

Maersk is shaping the future of maritime by embracing sustainability and tech innovation while fostering partnerships. They're not waiting for change; they're driving it.

When we talk about Maersk's future plans, we're talking about a company that's already operating in the future. They're steering the ship of innovation while keeping an eye on the planet's sustainability. It's this combination of ambition and responsibility that sets them apart.

MSC: The Titan's Journey

If we consider Maersk as the pioneer in the industry, then MSC is no less. It is a formidable force in its own regard.

Evolution of MSC over the Years

Think of 1970—no smartphones, no internet, but the birth of something big. That year, MSC began its journey. It started small, but make no mistake, this company had ambition written all over it.

Fast-forward to today. Currently, MSC is the world's second-largest shipping container carrier.

What catapulted MSC from a fledgling venture to a dominant player? In one word: **diversification**. Imagine a musician who doesn't just play the guitar but also drums, sings, and writes songs. MSC did just that. They went beyond shipping containers to venture into cruising, stretched their arms into logistics, and even sank their teeth into terminal investments.

But diversification wasn't just a strategy to make more money. It was a survival tactic. In an industry where fortunes can flip overnight due to economic uncertainties, diversification made MSC resilient. They weren't merely playing the game; they were changing it. They leveraged what they were good at, used it as a stepping stone, and ventured into uncharted waters—literally and figuratively.

MSC used its initial success in shipping as a launchpad. From there, they shot up like a rocket, reaching for the stars in various sectors. And this wasn't by accident. It was a calculated move designed to ensure long-term stability while setting the stage for growth and innovation.

Pioneering Initiatives and Aspirations for the Future

Taking a break is not in line with MSC's approach. The primary objective of this organization is oriented toward future-oriented endeavors. They're making industry-changing reforms rather than mere incremental adjustments. Let's discuss.

Sustainability

Firstly, sustainability isn't just a catchphrase for MSC; it's a commitment. They're investing real money and effort into making their fleet eco-friendlier. How? By designing ships that have a smaller carbon footprint and by exploring fuels that are less harmful to the environment. That's not just good for PR; it's good for the planet. And in an age of climate change, that's not just optional; it's essential.

Technology

Some companies are still getting used to emails, and MSC is diving headfirst into artificial intelligence and machine learning. Their goal is to streamline their operations. From optimizing shipping routes to predicting maintenance needs before something breaks, technology is MSC's silent co-pilot, steering them toward greater efficiency.

But here's where MSC truly stands out: their vision. They're not looking at where the maritime industry will be next year or even five years from now. They're looking at the next decade and beyond. They're researching future shipping solutions, from automation to zero-emission technologies, that could change not only their business but the entire maritime industry.

MSC is content in not just reacting to changes; they're proactively driving them. They're evolving traditional shipping into something more sustainable, more efficient, and more exciting.

To sum it up, MSC isn't just a company that's survived in a tough, competitive industry; it's thrived and is now setting the agenda for the future. Just like their competitor, Maersk, MSC is a company that defines what modern maritime success looks like.

Other Key Players: The Rising Stars and the Established Giants

Since we have explored A-listers like Maersk and MSC, let us have a quick look at the other leading maritime players. The maritime industry is not a two-horse race, after all. There are rising stars making their mark and established giants who have been holding their ground for years.

CMA CGM

CMA CGM is another mammoth but one that has specialized in certain trade routes. It's like the expert known for a particular genre but nails it every time.

COSCO

Then there's COSCO, the Chinese giant. If MSC is the Jay-Z of maritime, think of COSCO as the Drake—a different style but equally impactful.

Hapag-Lloyd

Let's not forget the industry veterans like Hapag-Lloyd, who have been around for a good while. They're like the Meryl Streeps of the maritime world—consistent, reliable, and evergreen.

These companies might not be making headlines every other day, but their steady, solid performance keeps the industry ticking.

Strategies and Contributions of Emerging Disruptors

Now, let us redirect our attention towards the emerging entities in the field, commonly referred to as disruptors.

Flexport

In contrast to the older generation, which came up in an era with less technology, Flexport was born in the internet age and thrives on data analytics and software solutions. Flexport, despite being a smaller enterprise, should not be underestimated as it endeavors to redefine the established norms rather than simply participating in the existing industry practices.

Freightos

Another exciting name to watch is Freightos, a marketplace for logistics that brings transparency to an often-opaque industry. Think of it as the eBay of shipping, bringing buyers and sellers together in a straightforward way that was not possible before.

Many of these emerging disruptors are doubling down on sustainability and tech innovation. They might not have the size and scale of the giants, but they have the agility and the will to experiment.

Regional vs. Global Trends in Maritime

In the maritime industry, it is simple to get swept up in the global trends, but there are also distinct regional differences. How do these regional characteristics blend with global changes? Let's commence.

The Universal Factors: Influences Affecting All

First off, let's talk about what everyone in the maritime world, regardless of region, is experiencing.

Technological Advancements: A Common Tide

In the maritime industry, technological advancements are like a rising tide that lifts all boats. This is a global phenomenon that's changing the game for everyone, whether you're in bustling Asian ports like Singapore or navigating through the historic Mediterranean waters.

Artificial Intelligence

Firstly, let's talk about Artificial Intelligence, or AI for short. This isn't just a buzzword; it's a transformative force. Imagine this: instead of conducting time-consuming, manual checks for ship maintenance, AI algorithms can now predict when a part is likely to fail. That means less downtime and more efficiency. And the best part? This predictive maintenance tech is spreading worldwide, making ships smarter and operations smoother, no matter where you're located.

Blockchain Technology

But it's not just AI that's making waves. Blockchain technology is coming into the fray, offering unparalleled security and transparency in transactions. Gone are the days of paper-based, error-prone processes. With blockchain, everything from contracts to cargo tracking becomes tamper-proof and transparent. And guess what? This technology is not restricted to any particular area; it's gaining traction from east to west.

Drones

Let's not forget about drones. These flying gadgets might have started as consumer toys, but today, they're valuable business tools. Need to inspect the farthest reaches of a massive cargo ship? Send a drone. Need aerial views for better navigational planning? A drone can provide that, too. Drones are cutting costs and saving time in ports and ships globally.

Automation

If you think self-driving cars are great, wait till you hear about automated cranes and autonomous port vehicles. These aren't scenes from a futuristic movie; they're current realities in many modern ports. This sort of automation is revolutionizing labor practices and operational efficiencies. From automated cargo handling in Europe to robotic cranes in Asian ports, automation is a global trend that's redefining the very nuts and bolts of maritime operations.

Green Initiatives and Sustainability: A Global Priority

In today's world, "going green" is no longer a catchphrase; it's an urgent call for action that echoes across the maritime industry. As the ramifications of climate change become increasingly stark, sustainability moves from the realm of optional to obligatory. This isn't a trend localized to any particular region; it's a global cry.

First off, let's look at cleaner technologies, a front where both small and large companies are making strides. For instance, in Europe, we're seeing a wave of electric ships that run on batteries, reducing both air and noise pollution. These vessels don't just make environmental sense; they're also a nod to evolving regulations and public sentiments.

Across the globe in Asia, solar-powered ports are emerging as hotbeds of innovation. Imagine ports where the operations—from cargo handling to lighting—are powered by renewable solar energy. It's not a distant dream; it's a reality that's already setting standards for others to follow.

But this isn't just about individual initiatives; there's a larger regulatory framework that's driving the maritime industry towards greener pastures. Take the IMO 2020 sulfur cap as a case in point. This global regulation mandates that ships must use fuel with a sulfur content of 0.5% or less, dramatically down from the previous limit of 3.5%. Compliance isn't just a matter of following the law; it's about staying in business. Those who don't adapt risk not just fines but also reputational damage.

Why does this matter? Because it signifies that clean operations aren't just a moral imperative but a legal one, too. Companies aren't just adopting sustainable practices because it's the right thing to do (although it is); they're doing it because it's now a business necessity.

In summary, the green initiatives shaping the maritime industry are both a universal mandate and a competitive imperative. Fueled by a mix of regulatory push and technological pull, these initiatives are redefining what it means to be a responsible, forward-looking player in the maritime world.

Whether it is about managing a shipping line, operating a port, or strategizing for a maritime startup, know this: the global focus on sustainability is not a passing wave but a permanent shift. It's a movement that transcends borders and is reshaping the industry into one that's not just more profitable but also more sustainable. And that's a tide that lifts everyone.

Distinguishing Regional Variations

So, we've discussed the global tides in technology and sustainability affecting everyone. Let's zoom in a little closer. It's time to look at the regional variations that add a local savor to the maritime industry.

Factors Driving Regional Maritime Trends

Not all maritime waters are created equal, and the same goes for the ports that populate them. Different regions have their unique characteristics, whether it's the bustling trade hubs in Asia or the highly regulated waters in Europe.

In Asia, for instance, growth is often fueled by the rapid rise of consumer markets and manufacturing hubs. Ports like Shanghai and Singapore have become global giants partly because of this regional boom. It's the local version of "supply and demand" playing out on a massive scale.

On the other side of the world, European ports are leading the charge in environmental regulations and sustainability. What's driving this? It's a mix of government policies, public sentiment, and a rich history of maritime trade that values long-term stability over quick profits.

And let's not forget regions like Africa and Latin America. While they may not yet be as large as their counterparts in Asia and Europe, they offer unique opportunities for growth, especially as global trade routes diversify. In these regions, investment in infrastructure and skills training are pivotal trends shaping the local maritime sector.

Region-specific Challenges and Opportunities

Every region comes with its own set of challenges and opportunities, and understanding these is key for any maritime player.

North America

In North America, for example, the challenge often lies in aging infrastructure. While technological advancements can mitigate some of these issues, substantial investment in port facilities is a pressing need.

Middle East

On another side of the world, let's say the Middle East, the advanced, state-of-the-art ports like Jebel Ali in the United Arab Emirates have set the standard for port facilities. However, the geopolitical complexities of the area present a different and kind of complex type of challenge, requiring an intricate approach to navigate successfully. Despite the difficulties, there are also chances for great opportunities in this region.

Africa

Africa is rich in natural resources. This continent is also blessed with an expanding middle class. The opportunity for maritime growth here is immense, especially for companies that are willing to invest in long-term partnerships and infrastructure development.

In summary, although the maritime industry is bound by certain universal trends—think technology and sustainability—it also manifests in unique ways across different regions. Understanding these regional distinctions is critical for anyone looking to succeed in this complicated and dynamic industry. Whether it's leveraging opportunities in emerging markets or navigating challenges in more established ones, a localized understanding can be just as crucial as a global perspective.

Real-world Scenarios and Practical Examples

We have covered the theory, the global trends, and the regional sectors. Now, let's get our feet wet with some real-world examples.

Case Studies: Success Stories from Diverse Regions

EUROPE

Let's start with the Port of Rotterdam in Europe, a shining example of how embracing technology can revamp operations. The port uses an Internet of Things (IoT) platform to gather data on water and weather conditions, ship movements, and dock availability. What's the big deal? Well, this data-driven approach boosts efficiency, safety, and sustainability. It's a win-win-win.

Asia

Specifically, the Port of Singapore, where they've nailed down what it means to be a "smart port." They've integrated automated systems, like electric cranes and drones for security surveillance, into everyday operations. The next big stop for them is their investment in training programs to upskill the workforce. As a result, they are becoming a port that's unconventional and is prepared for future challenges.

Kenya

Innovation isn't just limited to the big players in the industry. The Port of Mombasa in Kenya might not be as large as Rotterdam or Singapore, but it's a prime example of how strategic investment can fuel growth. With upgrades in infrastructure and enhanced security measures, the port has managed to boost its cargo-handling capacity and attract more international trade.

Lessons from Regional Maritime Developments

What can we learn from these case studies?

Technology isn't a luxury; it's a necessity. Only sticking to the current situations and circumstances did not make Ports like Rotterdam and Singapore industry leaders. Instead, they invested in the latest tech, which paid off big time.

Do not underestimate the power of human capital. Singapore's emphasis on worker training underscores the significance of adequately preparing for a future that is technology-driven. This approach ensures that automation and artificial intelligence (AI) serve to enhance human talents rather than supplant them.

Lastly, size isn't everything. The Port of Mombasa proves that targeted, thoughtful investment can yield significant returns, no matter your starting point. The size of an enterprise is not the determining factor of its success; rather, it is the strategic and intelligent execution of its operations.

Employing effective techniques and making appropriate investments, companies possessing the necessary motivation, perseverance, and financial means can successfully navigate challenging circumstances and achieve their desired goals.

The Interwoven World of Maritime Transport

Now that we have had a closer look at the global trends, regional specifics, and some real-world examples, this is time to bring all these pieces together to see the bigger picture.

Merging the Global and Regional Perspectives

In maritime transport, the global and regional sectors aren't two separate entities; they're more like two sides of the same coin. Imagine this: a technological innovation born in a European port could solve an operational challenge in an Asian harbor. Likewise, a sustainability initiative piloted in a small African port could inspire changes in a sprawling American shipping hub.

Do you understand where we're getting at? Global trends often set the tone, but it's the regional dynamics that add color and texture. Like in a good recipe, each ingredient—whether it's a dash of local policy or a sprinkle of international regulation—contributes to the final flavor. Understanding this relationship is crucial. It helps maritime professionals to adapt to changes and seize opportunities that might not be immediately obvious.

For example, the global push for sustainability might inspire a port in Latin America to invest in solar energy. Meanwhile, the technological advancements popular in Asian ports could motivate European harbors to explore AI-based solutions. Its consequences are a more efficient, more sustainable, and ultimately more profitable maritime industry worldwide.

Anticipating the Future: A Brief Look Ahead

We've now examined our current situation. Now, let us assess the direction in which we are progressing. Regarding something as fast-paced as marine transportation, the future isn't far out in the distance; it's right around the bend.

The trends are clear: sustainability is not just a trend but a business imperative. Automation and digitalization are not optional; they're integral. The future will be shaped by those who can successfully integrate these global trends with regional insights.

Think about the future impact of AI and machine learning. The advancements in these technologies are leading to increased intricacy and intelligence, hence enhancing the accuracy of predictive analytics and real-time decision-making. This implies that regardless of whether an individual is a shipowner or port operator, adopting these technologies at an opportune moment will open the path for success.

But let's not overlook the importance of human skills. As technology advances, the need for specialized skills—think cybersecurity, data analysis, and environmental engineering—will only grow. Investing in workforce development today could be your ticket to a more competitive position tomorrow.

In conclusion, this chapter states that the maritime world is complicated. It is a conjunction of global trends and regional variations. The key to success in this industry lies in understanding the above concepts and knowing the art of making them work for you. The only constant in this industry is change. Preparation today is the best way to ensure smooth sailing in the future.

Chapter 2

THE EVOLUTION OF
VESSELS: DESIGNS OF TOMORROW

The maritime transportation system serves as the vital oceanic circulatory system of the global economy. When the term "maritime industry" is mentioned, it typically evokes images of colossal vessels traversing vast bodies of water. This pattern is comprehensible since ships represent an essential part of the maritime equation rather than being a simply peripheral component. These vessels play a significant role in the extensive narrative of globalization, international trade, and technical advancement, often without receiving due recognition.

These ships are more than just steel boxes on the ocean; they're the product of brilliant engineering and the nautical sciences. They have highly complex ecosystems. From massive commercial ships that travel continents to fishing boats that support coastal villages, ships are diverse and necessary.

You might wonder why it matters how these ships evolved over time. Primarily, these modifications mirror wider alterations in the fields of economics, technology, and geopolitics. Like a reflection, vessels represent a period's technological sophistication, commercial objectives, and cultural dynamics. As we move toward a more linked world, these vessels face significant challenges as well as possibilities. Various factors, including climate change, cyber risks, and disruptive technologies, significantly impact the construction, operation, and management of ships.

To put it briefly, this chapter seeks to provide you with an overview of the world of vessels. It will trace their history, explore their design and operating principles, and speculate on their future. Rather than only observing the horizon, we will employ a telescope to discern potential occurrences approaching our vicinity. As we navigate through the past, stay focused on the present, and plan for the future, remember that every word is navigational on our voyage.

The Rationale for Focusing on Vessel Evolution

Picture this: a serene waterfront where an ancient mariner paddles a simple wooden canoe. Now, flash forward to today, where giant ships, almost like moving cities, cruise through vast oceans, their powerful engines roaring. What is the difference between these two scenes? It's all about the evolution of vessels.

The story of ships is really the story of us—people. As we grew, learned, and changed, so did our boats and ships. Every new design, from sailboats to steamships and now to the green vessels of today, tells a tale about what mattered to us at that moment. Were we exploring new lands? Were we trying to trade faster? Or are we now, more than ever, concerned about our planet?

When we talk about the maritime industry's past and future, we're really zooming in on the ships. They're like the main characters in our story. They reflect the dreams, needs, and challenges of their times. And just like any main character, they've had their ups and downs. They've faced wars, economic downturns, and, yes, even the pressing demand to be more 'green.'

However, it is imperative to comprehend the historical and future trajectories of ships. This provides an opportunity to gain insights into potential future challenges that the marine industry may encounter. The key to addressing several challenges in the maritime industry, such as developing sustainable fuels, creating autonomous ships, and enhancing vessel safety in adverse weather conditions, lies in understanding the ship's journey.

So, as we set sail into this topic, remember: we're not just talking about ships; we're talking about human ambition, innovation, and the ever-present drive to do better.

Historical Perspective: The Changing Face of Vessels

The vast oceans of maritime history reveal an intriguing relationship between human advancement and aspiration, with ships serving as physical symbols of our voyage. The story of these ships gives us a fascinating look into our shared hopes and dreams, echoing the spirit of the time they were made. Let's embark on this illuminating voyage through the revolutionary eras of shipbuilding and design.

From Wooden Boats to Steel Giants

In the beginning, our ancestors used to craft boats from logs and timber. These wooden vessels, though seemingly simple, were marvels of their time, allowing early humans to explore unknown waters, connect distant shores, and pave the way for intercultural exchanges. These boats were more than just the art of tools; they embodied the audacity and curiosity of humans willing to venture into the vast unknown.

During the wooden age, a diverse range of designs emerged, encompassing a variety of watercraft, such as single-log canoes and multi-mast sailing ships. These adaptations were specifically tailored to fulfill distinct purposes, including fishing, trading, and exploration. Nevertheless, with the expansion of human cultures and the increasing complexity of global commerce networks, there emerged a necessity for vessels capable of accommodating larger cargo loads, traversing longer distances, and withstanding the most challenging maritime environments.

Then, a new era of steel commenced. The integration of steel into the construction of ships marked a significant transformation in the field of maritime engineering. Shipbuilders were no longer constrained by the restrictions imposed by timber, therefore enabling them to envision and pursue more ambitious and audacious designs. The introduction of steel into maritime construction offered enhanced durability, stability, and the potential for more ambitious expeditions. The evolution of ships involved a shift from their utilitarian function as simply carriers to becoming formidable structures that symbolized power and status. The advent of steel also represented a notable transition in naval warfare, precipitating the emergence of battleships and cruisers that achieved supremacy over maritime domains.

Beyond their practical advantages, these steel behemoths became symbols of national pride and technological prowess. Ports, once quaint docking points, transformed into bustling hubs of activity, equipped to service the needs of these majestic giants.

The Age of Steam and Coal

The whispers of change began with the cogs and wheels of the Industrial Revolution. As factories started dotting the outlook and machinery took center stage, the maritime world found itself on the cusp of a monumental transformation.

Gone were the days when ships solely depended on the unpredictable whims of the wind. With the advent of the steam engine, vessels acquired a newfound autonomy. This was more than just a shift in propulsion; it was a leap into a new era of possibilities. The once-silent seas began to reverberate with the rhythmic chugging of steamships, symbols of human ingenuity and determination.

Fueling this steam revolution was coal. Dark, dusty, and dense, this mineral became the driving force behind the world's newfound obsession with steam power. It wasn't just about burning coal to produce steam; it was about harnessing that power to propel massive steel structures across vast oceanic distances. Ports transformed from quiet docking points to bustling hubs of commerce, with coal at the heart of this frenzy. The sight of coal-laden ships and the sounds of shovels and cranes became synonymous with progress.

Not only did the steam and coal era see unprecedented invention, but it also saw faster ships. Engineers, scientists, and shipbuilders engaged in collaborative efforts, conducted experiments, and frequently undertook daring ventures. The designs of vessels underwent a process of evolution, with a primary emphasis on optimizing efficiency, stability, and cargo capacity. The implementation of safety procedures, navigation aids, and communication systems resulted in enhanced security and predictability during maritime excursions.

Several of the ships during the time period achieved legendary status. The narratives, encompassing opulence, velocity, and even calamity, captivated the collective consciousness of individuals across the globe. The ships in question transcended their utilitarian purpose and instead symbolized the ambitions, aspirations, and, at times, the excessive pride of the era.

The Age of Steam and Coal shows humanity's constant pursuit of progress. It reminds us of a time when steam and steel fueled miniaturization, surmountable limits, and opportunity-filled oceans.

The Transition to Modern Container Ships

Consider, for a moment, a hypothetical scenario in which various commodities, ranging from rare spices to advanced electrical devices, were systematically and slowly transferred onto maritime vessels using manual handling processes. The task required significant manual effort, consumed a considerable amount of time, and frequently exhibited inefficiencies. With the increasing speed of global trade, the maritime industry acknowledged the necessity for a more efficient and simplified method. The period of containerization emerged as a revolutionary transformation that had a lasting impact on the context of international transport.

With the rise of mechanized vessels and an ever-globalizing world economy, the maritime sector had to adapt fast. The solution? Standardized containers. These weren't just metal boxes but symbols of a new age of efficiency. By introducing a uniform system, goods could be packed in factories and warehouses and loaded directly onto ships, trains, or trucks. The seamless transfer reduced the chances of damage, theft, and delays, marking a significant leap in shipping efficiency.

But the containers were just one part of the puzzle. Ships themselves underwent a transformative redesign. The bulky vessels of the past made way for sleek, streamlined ships explicitly designed to carry these containers. Stacked high and loaded efficiently, these container ships became the workhorses of global trade. They were designed to maximize capacity without compromising on speed or safety.

As these container ships grew in size and number, ports around the world had to adapt. The quiet docks of yesteryears morphed into bustling terminals equipped with towering cranes, intricate conveyor systems, and advanced logistics capabilities. These ports became logistical masterpieces, capable of handling the mammoth task of loading and unloading these container-laden giants.

In reflecting on this transformative period, it becomes evident that the shift to modern container ships was more than just a logistical decision. It represented a broader vision of a connected world, where goods could move seamlessly across oceans and continents. It's a testament to the maritime industry's adaptability and foresight, ensuring that as the world moved forward, so did the ships that connected it.

Modern Trends in Ship Design and Construction

As we progress through the contemporary era, the marine sector is characterized by a fusion of artistic expression, technological innovations, and dynamic international norms. The vessels traversing the world's waters in contemporary times serve a purpose traverse the world's waters in contemporary times serve a purpose that extends beyond mere transportation. They represent the realization of human aspirations, our increasing awareness of our technological prowess, and our developing sense of environmental responsibility.

In our modern day, the practice of shipbuilding extends beyond the mere construction of a maritime vessel, encompassing the complex procedure of conceptualizing and materializing a visionary creation. Modern vessels are meticulously engineered, considering a complicated equilibrium between efficiency and environmental responsibility. The equipment is furnished with cutting-edge technology, thereby guaranteeing safety, effectiveness, and limited environmental repercussions.

But it's not just about the ships themselves. The methods and materials used in their construction have also seen revolutionary changes, embracing both innovation and tradition. From the use of advanced, eco-friendly materials to digital design simulations that predict a ship's performance long before it hits the water, the entire process of shipbuilding has been transformed.

Moreover, in an interconnected world, the maritime industry doesn't operate in a vacuum. It is influenced by global events, market demands, and, significantly, international regulations. The push for cleaner, greener vessels, for instance, is not just a result of technological advancements but also a response to global calls for sustainability and environmental preservation.

Contemporary Ship Designs and Their Limitations

Within the broader context of maritime development, contemporary ship designs epitomize the pinnacle of human ingenuity and technical expertise. The modern vessels, encompassing both efficient commercial carriers and luxurious cruise ships, exemplify exceptional craftsmanship by seamlessly integrating aesthetic design and practicality.

Modern ship designs prioritize multiple aspects: efficiency to ensure optimal fuel consumption and voyage speed; safety to safeguard the crew, passengers, and cargo; and adaptability, allowing these vessels to serve various roles in the ever-evolving maritime sector. The use of advanced materials, such as lightweight composites combined with age-old steel, enhances structural integrity. The incorporation of aerodynamic principles reduces drag, and modular construction offers flexibility, making these ships agile and responsive.

Yet, like all feats of innovation, modern ship designs come with their own set of challenges. One of the most prominent is the sheer size of some of these vessels. The quest for efficiency has led to the creation of ultra-large ships capable of carrying vast amounts of cargo or thousands of passengers. While their size presents clear economic advantages, it also introduces logistical constraints. Many ports around the world aren't equipped to accommodate these giants, necessitating the construction of specialized deep-water terminals or limiting the vessel's accessibility to specific ports.

Furthermore, the magnitude of these ships brings forth navigational challenges. Maneuvering such large vessels, especially in narrow channels or busy ports, requires exceptional skill and precision. The risk factors increase, with the potential for significant environmental and financial consequences in the event of mishaps.

Modern ships are equipped with complex onboard systems that are meant to simplify and manage operations. Although automation and efficiency are notable advantages, they necessitate a proficient crew with a comprehensive understanding of contemporary nautical technology. The prioritization of training is of utmost importance, as it guarantees that sailors possess the necessary skills to effectively operate advanced equipment and address any potential technological malfunctions.

Essentially, contemporary ship designs exemplify the ultimate level of maritime accomplishments, hence emphasizing the significance of ongoing education, adaptability, and forward-looking perspectives. The industry's ability to strike a balance between innovation and pragmatism will play a pivotal role in determining its direction in the future.

Technological Advancements in Shipbuilding

At the core of the shipbuilding industry is an innovative spirit that constantly pushes the limits of possibilities. In the contemporary period, the general consensus is increasingly apparent, mainly due to the rapid progress in technology, which has brought about a prosperous period for both shipbuilders and mariners.

Contemporary shipbuilding practices encompass more than traditional tools and materials such as hammers, nails, and steel. Instead, they involve strategically utilizing state-of-the-art technology to reevaluate the fundamental nature of maritime construction. Digital design tools, for instance, offer shipbuilders a canvas to sketch their visions. These tools, complemented by advanced simulation software, allow for meticulous planning, ensuring every design detail is optimized even before the first piece of metal is shaped. The result? Ships that are not only robust and efficient but also tailored to the specific needs of the maritime world.

3D printing, once a concept restricted to science fiction, is now a tangible reality in shipyards. This technology has opened up new horizons in manufacturing ship components. Whether it's crafting a bespoke part for a luxury liner or producing large-scale components with pinpoint accuracy, 3D printing is reducing waste, speeding up production, and offering unparalleled customization.

However, the technological symphony doesn't stop at design and manufacturing. Once these ships set sail, they are veritable hubs of technology. Equipped with a myriad of sensors, they constantly gather data, from the ship's performance metrics to environmental conditions. Integrated automation systems ensure smooth operations, reducing human error and optimizing processes. Advanced data analytics tools process this wealth of information, offering insights that can enhance operational efficiency, bolster navigation accuracy, and fortify safety protocols.

A noteworthy trend in the maritime setting is the industry's unwavering focus on sustainability. As global conversations pivot towards environmental conservation, shipbuilding is no exception. Electric drives, biofuels, and hybrid engines show the industry's commitment to a greener future. Lowering operational expenses, these ecologically sustainable technologies mitigate the effects of carbon emissions and present possible long-term economic advantages.

As shipbuilding innovates and adapts, it envisions a future where marine activities are efficient, environmentally responsible, and in line with global needs.

The Role of Regulations and Standards

When we navigate the vast oceans of maritime, it becomes apparent that ships aren't just influenced by the winds and tides but also by the challenging web of regulations and standards set by international bodies. These regulations serve as guiding stars, ensuring the industry doesn't lose its way amidst the challenges of our rapidly changing world.

In today's interconnected global society, the maritime sector plays an instrumental role in linking economies, cultures, and ecosystems. But this vast network comes with its own set of responsibilities. As concerns over environmental conservation and safety mount, the industry finds itself under the watchful eyes of regulatory bodies. International and national institutions guarantee that maritime operations follow best practices to protect the planet and people.

At the core of the regulatory framework lies the International Maritime Organization (IMO). Acting as the maritime world's guardian, the IMO ensures that ships don't just sail but do so responsibly. The IMO's targets, especially in the area of greenhouse gas emissions, are ambitious and indicative of a larger vision for a sustainable future. In addition to establishing a benchmark, the IMO is pushing the industry to innovate by promoting a large decrease in emissions.

These regulations have a ripple effect, influencing various facets of the maritime sector. Shipbuilders, for instance, are now exploring eco-friendly materials and designs that reduce environmental impact. Operators are being pushed to think beyond traditional fuels, exploring alternatives like biofuels, hydrogen, and electric propulsion. Ports, too, are reimagining their operations, integrating green technologies and waste management practices to minimize their ecological footprint.

However, it is important to note that the scope of concern extends beyond merely the environment. Safety laws are implemented to guarantee that ships are designed and operated with the highest level of consideration for the preservation of human life. From the design of escape routes to the quality of onboard equipment, every detail is scrutinized to ensure the safety of both the crew and the cargo.

Sustainability: The Green Revolution in Ship Design

The maritime industry is undergoing a vibrant transformation, tinted with shades of green. As global discussions amplify around environmental conservation, the maritime industry is steering its course toward a more sustainable future. The ships of today and tomorrow are not just reflections of engineering prowess but are emblematic of a broader commitment to the planet. Let's closely look into the waves of this green revolution.

Importance of Eco-friendly Designs:

In an era characterized by heightened awareness of the environmental impacts of human activities, the development and implementation of eco-friendly ship designs have evolved from mere ideals to becoming necessary demands. The marine sector plays a substantial role in facilitating international trade and is pivotal in addressing and minimizing environmental consequences.

Why the emphasis on eco-friendly designs? First and foremost, traditional ship operations contribute substantially to global emissions. As the world constricts its carbon budget, reducing emissions from ships becomes paramount. Furthermore, oceans and waterways, the very lifelines of ships, are vulnerable ecosystems. Oil spills, waste disposal, and noise pollution can have cascading impacts on marine life.

Incorporating sustainable designs means not only reducing the carbon footprint but also ensuring that ships harmoniously coexist with the environment. Such designs can lead to economic benefits, too, as fuel-efficient ships translate to lower operational costs.

Examples of Sustainable Technologies

Several pioneering technologies are at the forefront of the green revolution in ship design:

Alternative Fuels: Traditional fossil fuels are making way for biofuels, liquefied natural gas (LNG), and hydrogen. These fuels not only reduce emissions but also optimize engine performance.

Solar and Wind Power Integration: Harnessing renewable energy, ships are now being equipped with solar panels and retractable wind sails, reducing their dependence on traditional fuels.

Waste Heat Recovery: Modern ships are being designed to capture and reuse waste heat from engines, turning it into an energy source and enhancing overall efficiency.

Ballast Water Treatment Systems: To prevent the transfer of harmful aquatic organisms, ships are adopting systems that treat ballast water before it's released.

Advanced Hull Designs: Aerodynamic hulls, designed using sophisticated computational fluid dynamics software, reduce water resistance, enhancing fuel efficiency.

Future Outlook on Green Vessels

Looking ahead, the maritime canvas is set to be awash with more shades of green. As regulatory pressures increase and technological innovations continue, the industry will likely witness a surge in fully electric ships, especially for short-haul routes.

Research into nuclear propulsion, once considered a distant dream, is gaining momentum, promising zero-emission voyages. Automation and artificial intelligence will play pivotal roles in optimizing routes for fuel efficiency and minimizing environmental impact.

Moreover, the industry will likely see a rise in "circular shipbuilding" - an approach where ships are designed for recyclability, ensuring that at the end of their lifecycle, materials are repurposed, reducing waste.

The pursuit of sustainability within the marine industry exemplifies the shared aspiration of humanity for a more environmentally conscious future. As we move forward, we hope to see bluer sky and clearer oceans. The tides of change have already begun to roll in.

Anticipated Breakthroughs in Vessel Architecture

As the new era in maritime history approaches, vessel architecture stands at the confluence of tradition and innovation. The shipbuilding industry, revered for its age-old practices, is now gearing up for transformative changes that promise to redefine the very essence of vessel design. Let's set our sights on the horizon and explore the breakthroughs anticipated in vessel architecture.

Advanced Materials for Ship Construction

The field of shipbuilding, which is deeply rooted in longstanding customs and practices, is currently experiencing a notable shift, particularly in the selection of materials employed. In addition to the conventional criteria of strength and durability, there is an increasing focus on the incorporation of sustainability and adaptability to address contemporary technological needs. In contemporary times, ship materials are required to possess environmental sustainability, compatibility with integrated technology, as well as a combination of lightweight properties and structural durability. In summary, the materials utilized in the construction of contemporary vessels embody a synthesis of traditional knowledge and advanced technological advancements, thereby driving the marine sector towards a sustainable route.

Graphene: Touted as the material of the future, graphene, with its remarkable strength and lightweight properties, is poised to revolutionize ship construction. Being a good conductor it can also enhance onboard systems' efficiency.

Titanium Alloys: While traditionally reserved for specific components due to cost, advancements in production techniques might make titanium alloys more accessible. Their corrosion resistance and strength-to-weight ratio make them ideal for maritime applications.

Composite Materials: Combinations of fibers and resins result in materials that are both strong and lightweight. Their resistance to corrosion also means reduced maintenance costs and longer vessel lifespans.

Eco-friendly Paints and Coatings: Bio-based paints and coatings that deter marine organisms without harming the environment are being developed, reducing drag and enhancing fuel efficiency.

The Beginning of Autonomous Vessels

The idea of ships sailing without a crew might have been the stuff of science fiction a few decades ago, but not anymore. Here is how the maritime industry has advanced in this sector:

Sensor Technology: Advanced sensors, combined with AI, allow ships to perceive their surroundings in real-time, making autonomous navigation possible.

Onboard Artificial Intelligence: AI systems can predict equipment failures, optimize routes for fuel efficiency, and make real-time decisions in complex navigational scenarios.

Satellite Communication: Enhanced satellite communication ensures that even in the vast expanse of the oceans, autonomous vessels remain connected, relaying critical data and receiving navigational inputs.

Safety Protocols: While the move towards autonomy is exciting, safety remains paramount. Advanced collision avoidance systems and emergency response protocols are being developed to ensure autonomous operations are safe.

Incorporating Renewable Energy Sources

The maritime industry's commitment to a greener future is evident in its exploration of renewable energy sources, which are as follows:

Solar Integration: The use of photovoltaic cells, either embedded in a ship's structure or as additional panels, can supplement a vessel's power needs, especially for onboard systems.

Wind Propulsion: While sails are a thing of the past, modern innovations like retractable wind turbines and wing sails harness wind energy, providing an eco-friendly boost to propulsion.

Wave Energy: Pioneering technologies aim to convert the kinetic energy of ocean waves into electricity, offering a constant power source on the high seas.

Hybrid Systems: Combining traditional fuels with renewable sources, hybrid propulsion systems promise efficiency, reduced emissions, and adaptability to various operational needs.

The projected advancements in vessel construction indicate our inventiveness and our enduring connection with the waters. They also serve as a roadmap for shaping the future. As we near these developments, the maritime industry promises an interesting ride.

Case Study: Maersk's Pioneering Vessel Advancements

Within the expansive field of marine trade, few entities possess a reputation as renowned as Maersk. A trailblazer in vessel advancements, Maersk's journey offers invaluable insights into the confluence of tradition, innovation, and a steadfast commitment to excellence.

Background and History

Founded in 1904, Maersk's maritime journey began with a single steamship. Throughout the next century, the organization experienced significant growth in its fleet and operational capacities, thereby establishing itself as a highly renowned organization in the context of international shipping. Maersk has journeyed from its modest Danish beginnings into the annals of maritime history, representing tenacity, aspiration, and sharp insight to adjust to shifting global environments.

Innovations and Adopted Technologies

Maersk's commitment to innovation is evident in its vessels, which are often considered benchmarks in modern ship design. Some of them are:

Triple-E Vessels: Introduced in 2013, Maersk's Triple-E class vessels were game-changers. Designed for energy efficiency, environmental performance, and economies of scale, these ships set new standards for cargo capacity and fuel consumption.

Alternative Fuels: Maersk has been actively engaged in the pursuit of sustainability and is doing extensive research and development in the areas of biofuels, liquefied natural gas (LNG), and other environmentally friendly alternatives. The primary objective of these efforts is to achieve a substantial reduction in carbon emissions.

Digital Transformation: Maersk's ships are equipped with advanced digital systems, enabling real-time data monitoring, predictive maintenance, and enhanced operational efficiency. Their embrace of blockchain technology for cargo tracking underscores their commitment to technological advancements.

Recycling Initiatives: Emphasizing responsible shipbreaking, Maersk ensures its vessels, at the end of their operational lifecycle, are recycled in facilities that adhere to the Hong Kong Convention standards, prioritizing both environmental and worker safety.

Lessons Learned and Future Plans

Throughout its expedition, Maersk has learned the value of adaptability. By staying ahead of regulatory curves and investing in research and development, the company has been able to navigate challenges, from economic downturns to changing environmental regulations.

For the future, Maersk is charting a course towards complete carbon neutrality by 2050. They are investing in research on green propulsion technologies, looking into wind power solutions, and exploring fully electric vessels for short-haul routes. With a focus on sustainable growth, Maersk aims to reinforce its position as an industry leader, setting benchmarks not just in operations but also in environmental stewardship.

The growth of this company's vessels is more than just a business venture. It shows the maritime industry's potential when ambition, innovation, and responsibility sail toward a common goal.

The Role of R&D: Fueling Future Innovations

In the ever-evolving seascape of the maritime industry, the beacon guiding progress is undeniably Research and Development (R&D). It's the compass that charts unexplored waters, the wind that propels vessels into uncharted territories, and the anchor ensuring stability amidst the tides of change. Let's delve deeper into the indispensable role R&D plays in the maritime domain.

Importance of Research and Development

R&D, in essence, represents the heart and soul of innovation within the maritime sector. It's the crucible where ideas are born, tested, refined, and eventually transformed into groundbreaking solutions.

Meeting Global Demands: As global trade expands and diversifies, there's a pressing need for vessels that are more efficient, faster, and adaptable. R&D endeavors strive to meet these demands, ensuring the maritime industry remains responsive to global needs.

Environmental Responsibility: In an era marked by environmental concerns, R&D initiatives focus on developing eco-friendly technologies, from alternative fuels to waste management systems, ensuring maritime activities align with global sustainability goals.

Safety Enhancements: The maritime industry, given its inherent risks, requires constant advancements in safety protocols and equipment. R&D plays a pivotal role in this, introducing innovations that safeguard both crew and cargo.

Case Examples of Successful R&D Initiatives

Ballast-Free Ship Design

For centuries, ballast systems have been the unsung heroes of maritime voyages, ensuring ships maintain their stability and equilibrium. Traditional ballast tanks, which use seawater to balance ships, inadvertently also become carriers for marine microorganisms. When this ballast water is released into foreign ecosystems, it can introduce invasive species, disrupting local marine environments and often leading to ecological imbalances.

Recognizing the environmental implications, R&D teams embarked on a quest to rethink ship stability. The result? The innovative ballast-free ship design. By altering the ship's structure and utilizing flow-through designs, these vessels maintain their stability without the need for holding and discharging ballast water. This ingenious design not only reduces the risk of introducing invasive marine species but also signifies the maritime industry's commitment to safeguarding marine ecosystems.

Air Lubrication Systems

In the relentless chase for efficiency, researchers turned their focus to a ship's hull, a primary point of resistance as vessels slice through the waters. Through extensive R&D, the concept of air lubrication systems was born. These systems release a steady stream of microbubbles along the ship's hull. This carpet of bubbles acts as a lubricating layer, significantly reducing water resistance.

The implications are profound. With decreased friction, ships can glide more smoothly, leading to enhanced fuel efficiency and reduced carbon emissions. Beyond the environmental benefits, this innovation translates to tangible economic savings for operators, showcasing how environmental responsibility and economic prudence can sail hand in hand.

Hydrogen Fuel Cells

As the global clarion call for cleaner energy grows louder, the maritime industry has been actively scouting for sustainable fuel alternatives. Enter hydrogen fuel cells, an R&D marvel that promises to redefine maritime propulsion. Unlike conventional fuels that release carbon and other pollutants, hydrogen fuel cells generate power through a chemical process, emitting only water vapor as a byproduct.

Integrating hydrogen fuel cells into ships is no small feat. It requires extensive modifications, from storage solutions for hydrogen, often in liquid form at extremely cold temperatures, to adapting engines to utilize this clean energy source. However, the rewards are immense. Vessels powered by hydrogen promise significant reductions in greenhouse gas emissions, making them pivotal players in the global push toward a greener future.

How R&D Will Shape the Vessels of the Future

The vessels of the future will be reflections of the R&D endeavors of today. As we gaze into the maritime future, we envision ships that are not just modes of transportation but also symbols of technological prowess and environmental harmony.

Fully Autonomous Vessels: Continued R&D will likely see the dream of fully autonomous ships become a reality, with advanced AI systems handling navigation, operations, and decision-making.

Modular Ship Designs: Future vessels might be modular, allowing for customization based on cargo type, journey duration, and environmental conditions.

Green Energy Hubs: Ships might not just utilize green energy but also produce and store it, becoming floating renewable energy hubs that can supply power to coastal regions.

R&D isn't just a theoretical exercise within the maritime industry. It's the lifeblood that promises progress, efficiency and a harmonious balance with our planet's needs.

Final Thoughts on the Future of Vessel Design

Our journey through the maritime industry's evolution underscores two pivotal themes: the relentless drive for innovation and an unwavering commitment to sustainability. From the humble beginnings of wooden boats to the dawn of autonomous vessels, the industry has continuously adapted, integrating cutting-edge technology and embracing eco-friendly practices. As we anchor our exploration, it's evident that the future of vessel design hinges on continued research and collaboration. Stakeholders must invest in R&D, share insights, and proactively engage with regulatory bodies. As we set sail towards this promising horizon, the vessels of tomorrow symbolize more than transportation – they represent hope, resilience, and a collective aspiration for a greener, interconnected world.

Fundamentally, the maritime sector is poised for a revolutionary period propelled by both technological progress and ecological consciousness. Through the collaborative efforts of all relevant parties involved, the industry is well-positioned to effectively address the obstacles that lie ahead and strategically pave the way toward a future characterized by sustainability and innovation.

Chapter 3

MANAGEMENT PARADIGMS: LEADING IN THE NEW MARITIME AGE

Traditional Management Structures and Their Evolution

The maritime world has always been governed by a sense of order, and much of it can be attributed to the deeply ingrained hierarchical system onboard ships. This structure's origins can be traced back to the earliest days of sailing when communication methods were limited and the unpredictability of the seas demanded a well-defined command chain.

Central to this hierarchy was the captain, a figure often draped in a cloak of authority and respect. The captain's decisions were more than just commands; they were the difference between safety and peril in the vast, often treacherous waters. His experience, intuition, and leadership ensured that the vessel stayed its course, even in the face of adversity.

Directly beneath the captain were the officers, each specializing in specific ship functions. The chief officer or first mate was responsible for the ship's overall management, the navigation officer charting the course, the chief engineer ensuring the ship's machinery ran seamlessly, and so on. These officers formed the bridge between the captain's directives and the crew's execution.

Then came the crew members, the heart and soul of the ship's daily operations. Ranging from deckhands, responsible for tasks like mooring and maintenance, to galley staff ensuring everyone onboard was well-fed, their roles, though varied, were unified in purpose: ensuring the ship's smooth sailing.

This clear delineation of roles and responsibilities, set against the backdrop of maritime tradition, ensured a cohesive operation. Everyone onboard knew their duties, whom to report to, and what to expect. It was a system that, for centuries, ensured that ships, despite facing the unpredictability of the oceans, remained oases of order and discipline.

The Shift: Reasons for Moving Away from Traditional Hierarchies:

The classic hierarchical model, with its rigid chain of command, was the backbone of maritime management for ages. It instilled a sense of order and discipline and ensured that decisions, often made under intense pressure, were swiftly executed. However, with the dawning of the modern maritime era, this age-old structure started showing signs of strain, prompting an industry-wide introspection. Here's a more profound understanding of the catalysts that precipitated this shift:

Technological Advancements

The maritime industry, once resistant to change, began embracing technology with open arms. As automation systems became more sophisticated and navigation tools more advanced, the traditional roles onboard ships started to evolve. Navigation officers now had automated systems to rely upon; engineers had machinery that could self-diagnose faults. This shift meant that a singular decision-making point was no longer efficient. Instead, a collaborative approach, where multiple crew members could provide inputs based on real-time data, became vital.

Globalization and Diverse Crews

The beauty of the modern maritime industry is its melting pot nature. Ships often boast crew members from various parts of the world, each bringing their cultural nuances, languages, and perspectives. While this diversity is a strength, a rigid hierarchical structure sometimes becomes an impediment. It sometimes hindered open communication and made mutual understanding challenging. For ships to

24

operate seamlessly, fostering a collaborative environment where every voice is heard and valued became crucial.

Need for Flexibility

The maritime world of today faces challenges its predecessors could scarcely imagine. From navigating the intricacies of geopolitics and dealing with piracy threats in certain maritime routes to maneuvering through increasingly congested waterways, modern-day challenges demand agility. A hierarchical structure, where every decision had to trickle down a chain, was ill-suited for such scenarios. What became essential was a nimble approach, where decision-making could be decentralized, allowing for swift, on-the-spot judgments.

In essence, while the traditional hierarchical structure served the maritime industry loyally for generations, the winds of change, brought about by technology, globalization, and modern-day challenges, necessitated a fresh approach. One that values collaboration, embraces diversity, and prizes flexibility, setting the maritime industry on a new course for the future.

Flexible Leadership Structures: Benefits and Implementations

The maritime industry, traditionally characterized by its rigid hierarchies, has been undergoing a sea change, moving towards more flexible leadership structures. This shift has not only been essential but also incredibly beneficial, reshaping the way vessels operate and fostering a more inclusive and efficient work environment.

Benefits

Enhanced Collaboration

The shift towards a more egalitarian structure has torn down the walls that once siloed crew members based on their rank or role. This has resulted in a more transparent and open communication channel where everyone, from the captain to the newest recruit, has a voice. It encourages diverse perspectives, drawing from the collective experience onboard, leading to more holistic solutions and fostering a genuine spirit of teamwork.

Empowerment and Morale Boost

In a flexible leadership structure, every crew member is recognized as a valuable asset, not just a cog in the machine. They're empowered to take initiative, suggest improvements, and even challenge traditional norms constructively. This sense of empowerment does wonders for morale, reducing turnover rates and fostering a culture where crew members are motivated to give their best.

Efficient Decision Making

While the captain retains the ultimate decision-making authority, a flexible structure ensures that decisions are made promptly, leveraging the collective intelligence onboard. In situations where quick judgment is paramount, like navigating through dense traffic or responding to emergencies, this collaborative decision-making process can be invaluable.

Implementations

Incorporating flexible leadership structures isn't about just discarding the old rulebook; it's about redefining roles, responsibilities, and processes:

Regular Team Meetings

These are platforms where everyone, irrespective of their rank, can share updates, voice concerns, and offer suggestions. It fosters a sense of collective ownership of the ship's operations and goals.

Collaborative Problem-Solving Sessions

Instead of a top-down approach, problems are discussed openly, drawing on the diverse experiences of crew members, leading to well-rounded solutions.

Cross-Training

To ensure a truly collaborative environment, crew members are often cross-trained in multiple roles. This ensures that everyone has a comprehensive understanding of the ship's operations, facilitating more effective teamwork and filling in operational gaps when needed.

The move towards flexible leadership structures in the maritime industry signifies a progressive step, one that acknowledges the collective strength of every individual onboard and harnesses it to navigate the challenges and opportunities of the modern maritime age.

Embracing Technological Innovations in Maritime Management
The Rise of Data-Driven Decisions in the Maritime Industry

In the marine industry, decision-making has traditionally relied heavily on extensive expertise, established procedures, and intuitive judgment. Nevertheless, the current state of the sector is experiencing a significant shift characterized by the emergence of data-driven decision-making.

Contemporary ships are outfitted with an advanced system of sensors that maintain constant surveillance over a multitude of factors. These sensors, strategically positioned, track everything from engine metrics and fuel efficiency to real-time weather conditions and cargo stability. This data, vast and comprehensive, offers a detailed snapshot of a vessel's operational status at any given moment.

The true power of this data is realized when it is synthesized and analyzed. Advanced data analytics tools, backed by robust algorithms, sift through this information, identifying trends, anomalies, and performance benchmarks. For ship operators and captains, this translates into actionable insights that inform their decisions. Instead of relying solely on past experiences, they now have empirical data to guide their choices, enhancing operational efficiency, ensuring safety, and optimizing costs.

Moreover, the integration of real-time data analytics allows for proactive measures. For instance, predictive maintenance can be scheduled based on actual equipment wear and tear rather than periodic schedules, preventing potential equipment failures and downtimes.

The maritime industry's shift towards data-driven decisions symbolizes its commitment to innovation, efficiency, and safety. As vessels continue to produce and utilize vast amounts of data, the industry stands poised to benefit from more informed, precise, and strategic decision-making.

The Role of Artificial Intelligence in Streamlining Maritime Operations

The marine sector is leading the way in adopting advanced technologies, with Artificial Intelligence (AI) serving as a notable example of this shift. The utilization of artificial intelligence (AI) has become crucial in contemporary maritime operations due to its growing intricacies and requirements. AI serves as a critical instrument in enhancing efficiency precision, and fostering innovative practices.

Predictive Maintenance

One of the most significant challenges in maritime operations is ensuring the optimal performance of machinery. Traditional maintenance schedules are often based on standard operational hours or calendar-based intervals. However, AI introduces predictive maintenance, analyzing data from machinery in real time to predict potential failures. By identifying wear and tear or anomalies in machine behavior, AI allows for timely intervention, minimizing downtimes and preventing costly repairs or replacements.

Optimal Route Planning

Navigating the vast oceans is no simple feat, especially with constantly changing weather patterns, sea conditions, and geopolitical scenarios. Artificial intelligence (AI) algorithms analyze extensive datasets, which encompass real-time satellite imagery, weather forecasts, and historical information, in order to provide recommendations for the most optimal and secure routes. In addition to conserving fuel and reducing travel time, this practice also serves to safeguard the well-being of both the crew members and the transported goods.

Automated Operations

The integration of AI extends beyond data analysis. Advanced systems now automate certain tasks aboard the ship, such as ballast water management or energy distribution, ensuring optimal performance. These systems can also learn from past operations, adapting and fine-tuning their actions for future voyages.

Safety and Surveillance

AI-powered surveillance systems enhance onboard safety. They can detect unauthorized personnel, monitor restricted areas, and even identify potential fire hazards or leaks. Furthermore, AI tools can analyze patterns and behaviors, flagging any anomalies that might indicate potential threats or issues.

Environmental Compliance

With stringent environmental regulations in place, ships need to ensure they minimize their ecological footprint. AI assists in monitoring emissions, waste disposal, and fuel consumption, ensuring that vessels adhere to global standards and minimize their environmental impact.

Incorporating Artificial Intelligence into marine operations is more than a simple improvement; rather, it signifies a fundamental transformation in approach. Through the utilization of artificial intelligence (AI), the maritime sector is positioned to effectively address the obstacles of the current era, displaying accuracy, efficacy, and proactive planning. This course of action establishes a path towards a forthcoming period that is not solely economically advantageous but also environmentally conscious and secure.

Big Data Insights: Predictive Analysis and Real-time Decision-Making

The contemporary maritime industry is in the middle of a digital transformation, with big data at its helm. The data generated by ships, ports, cargo, and even the oceanic environment itself, when harnessed effectively, can profoundly influence maritime operations.

Granularity of Data

Modern vessels are outfitted with numerous sensors and monitoring systems that continuously collect data. This includes information about engine performance, cargo conditions, fuel consumption, crew activities, and much more. The granularity of this data, when analyzed in aggregate, provides a comprehensive view of the ship's operations and its immediate environment.

Predictive Analysis

Leveraging historical data combined with current operational data, predictive analysis tools can forecast potential future events with remarkable accuracy. For instance, by analyzing past data on ocean currents, wind patterns, and shipping routes, systems can predict optimal paths for voyages. Similarly, by examining machinery performance trends, ships can predict potential malfunctions or breakdowns, leading to more timely maintenance and reduced downtimes.

Real-time Operational Adjustments

The true power of big data is realized when it's used for real-time decision-making. Advanced analytics tools process incoming data streams almost instantaneously. If a ship encounters an unpredicted storm front, real-time data analytics can provide alternative routes or suggest operational adjustments to minimize risks.

Data-driven Efficiency

Efficiency is at the core of maritime operations. By analyzing fuel consumption patterns, route efficiencies, and cargo management, big data tools can suggest measures to optimize fuel usage, reduce operational costs, and enhance cargo handling, ensuring that ships operate at peak efficiency.

Safety and Compliance Monitoring

Big data also plays a pivotal role in ensuring ships comply with international safety and environmental standards. Continuous monitoring and analysis of emissions, waste management, and safety protocols ensure that ships not only adhere to regulations but also identify areas for improvement.

Big data is not just an ancillary tool but an integral component in the modern maritime ecosystem. It offers the maritime industry a lens through which it can view its operations in unprecedented detail, making informed decisions that are not only reactive but also proactive. As the industry embraces big data, it charts a course toward an era characterized by enhanced efficiency, safety, and sustainability.

Case Studies: Successful Integration of Tech in Maritime Management

The maritime industry, historically perceived as slow to change, is now at the forefront of technological innovation. Companies are embracing advanced technologies not just as add-ons but as central pillars of their operational strategies. Here are some notable case studies highlighting the industry's technological strides.

Maersk Line's Remote Container Management System

Recognizing the challenges in managing perishable cargo during long voyages, Maersk Line initiated the Remote Container Management System, leveraging the power of the Internet of Things (IoT) and cloud technology.

Features and Benefits
Real-Time Monitoring
Equipped with IoT sensors, each refrigerated container transmits data regarding its internal conditions to a centralized cloud platform. This ensures that the temperature, humidity, and CO_2 levels are maintained within optimal ranges throughout the journey.

Reduction in Cargo Spoilage
By maintaining optimal conditions, Maersk has significantly reduced the spoilage of perishable goods, leading to increased customer satisfaction and reduced financial losses.

Data Analytics for Continuous Improvement
The data collected is analyzed to identify patterns and potential areas of improvement, paving the way for more efficient future operations and even predictive interventions.

MSC's Digital Twin Technology
Overview: Mediterranean Shipping Company (MSC), always a trailblazer, took a futuristic approach by adopting digital twin technology. This involves creating highly detailed digital replicas of their physical ships.

Features and Benefits Predictive Maintenance
By comparing real-time data from the ship with its digital twin, MSC can predict potential wear and tear on machinery and components. This foresight allows them to carry out maintenance during scheduled downtimes, avoiding unexpected breakdowns.

Operational Efficiency
The digital twin provides insights into the ship's performance, allowing for tweaks in operations, leading to improved fuel efficiency, speed optimizations, and overall better voyage planning.

Safety and Training
The digital twin can be used in simulation environments, providing a risk-free platform for crew training. It also offers a valuable tool for safety drills and understanding the ship's response to various emergency scenarios.

These case studies exemplify the maritime industry's commitment to integrating advanced technologies into its core operations. Companies like Maersk and MSC are not just riding the technological wave but are setting the course for others to follow. Their initiatives underscore the tangible benefits tech integration brings, from operational efficiencies and cost savings to enhanced safety and customer satisfaction.

Ethical Considerations in Modern Maritime Management
In an interconnected global economy, the maritime industry plays a pivotal role, and with that comes an assortment of ethical challenges and considerations. As ships bridge continents, they also navigate complex ethical waters, from environmental responsibilities to the treatment of crew members. This segment sheds light on the profound ethical imperatives shaping the maritime sector in today's age.

Understanding the Ethical Imperatives of Maritime Operations
The maritime industry, with its sprawling global footprint and intercontinental scope, is a pivotal player in the world's economy. With this vast reach comes a responsibility that transcends mere business objectives. The ethical imperatives in maritime operations form the backbone of responsible commerce, ensuring that the interests of all stakeholders, including the environment, are prioritized. These imperatives are more than just moral compasses—they translate into tangible outcomes that shape the very essence of maritime operations.

Significance

Trust and Reputation

Deepening Stakeholder Relationships
Trust isn't a given; it's earned. By upholding ethical standards, maritime organizations cultivate deeper, more enduring relationships with their stakeholders. This isn't limited to just clients and partners. It extends to communities where ports are located, regulatory bodies overseeing maritime operations, and the crew that forms the heart of every vessel.

Brand Perception

In an era in which information is readily available, a company's reputation is constantly scrutinized. Any breach of ethical standards can result in substantial reputational harm, affecting client relationships and the company's market value as well as its capacity to attract top talent.

Legal and Regulatory Compliance

Proactive Approach

Instead of reacting to regulations, ethical companies stay ahead, ensuring that their operations not only meet but often exceed the required standards. This proactiveness minimizes the risk of costly lawsuits, fines, and sanctions.

Global Standards

With ships often crossing multiple jurisdictions and international waters, adhering to a global standard of ethics ensures seamless operations. Such uniformity in ethical operations simplifies compliance procedures and fosters a universally accepted code of conduct.

Long-Term Viability Sustainable Growth

Ethical considerations prioritize long-term sustainability over short-term gains. By ensuring that operations don't deplete resources, whether human or environmental, maritime organizations pave the way for enduring growth. This translates to consistent returns, stakeholder loyalty, and a resilient business model that can weather economic downturns.

Mitigating Risks

Ethical operations are intrinsically aligned with risk mitigation. By avoiding practices that exploit vulnerabilities, be it in terms of environmental conservation or labor rights; maritime businesses prevent potential future liabilities. Such foresight not only ensures compliance but also shields the organization from reputational and financial setbacks.

Understanding and integrating ethical imperatives into maritime operations is not an optional endeavor—it's a foundational aspect that influences every facet of the maritime world, dictating its present actions and future trajectory.

Navigating Environmental Responsibilities and Sustainability

The maritime sector operates within the wide expanse of the oceans of the world, exerting influence on a significant portion of the global population and having effects on intricate marine ecosystems. Throughout history, the sector has been characterized by an unfavorable environmental track record, marked by significant emissions of greenhouse gases, occurrences of oil spills, and disturbances to marine environments. However, recognizing the imperative need for sustainability, there's been a notable shift towards environmentally responsible operations.

Approach

1. Reducing Emissions Shift to Alternative Fuels

Ships are gradually transitioning away from using traditional bunker fuels in favor of more environmentally friendly options, including liquified natural gas (LNG), biofuels, and even hydrogen fuel cells. When compared to conventional fuels, liquefied natural gas (LNG) significantly reduces carbon dioxide (CO_2) emissions by 25%. These alternate strategies have the potential to lessen carbon emissions and the discharge of sulfur and nitrogen oxides.

Energy-Efficient Ship Designs

Streamlined hull designs, improved propeller technologies, and air lubrication systems are being employed to reduce fuel consumption and, consequently, emissions. Additionally, the adoption of "cold ironing" allows ships to plug into shore-side electrical power, reducing the need to run onboard generators while docked.

2. Waste Management Advanced Treatment Systems

With strict international regulations like MARPOL Annex V, ships are now equipped with systems that segregate and treat waste onboard. This includes separating biodegradable waste from non-biodegradable materials and using advanced treatment systems for sewage and wastewater.

Recycling Initiatives

Many maritime companies have implemented robust recycling programs. Items, like used oil, batteries, and scrap metals, are stored and then offloaded at ports for proper recycling or disposal.

3. Protecting Marine Biodiversity

Ballast Water Management Systems (BWMS):

Recognizing the harm posed by invasive species carried in ballast water, the International Maritime Organization (IMO) has set guidelines that ships must follow. Modern ships are equipped with BWMS that treat ballast water, either through filtration, UV radiation, or chemical treatment, to eliminate foreign organisms before discharge.

Navigational Measures

Advanced navigational systems provide real-time data about marine habitats, allowing ships to avoid ecologically sensitive areas. This is particularly vital for protecting coral reefs, breeding grounds, and regions with high marine biodiversity.

Impact:

The proactive measures adopted by the maritime industry reflect a deep understanding of its environmental responsibilities. By innovating and adapting, the industry is not only meeting regulatory standards but is also setting a benchmark for sustainable operations in the broader transportation sector. More than compliance, this shift underscores the industry's commitment to preserving our planet for future generations.

Ethical Treatment of Crew: Welfare, Training, and Equal Opportunities

The maritime sector is built on the strength and resilience of its crew. These individuals brave the vast oceans, spending months away from their loved ones, ensuring smooth and efficient operations. As such, their well-being, continuous development, and equal treatment are not only ethically essential but also critical for the industry's sustainability and success.

Focus Areas:

1. Welfare Initiatives:

Health and Safety Protocols

Companies are investing in advanced medical facilities onboard, with trained medical personnel, ensuring the crew's health and well-being. Regular health check-ups, mental health support, and emergency medical procedures are increasingly becoming the norm.

Recreational and Communication Facilities

Recognizing the mental toll of long voyages, ships are equipped with recreational zones, including gyms, entertainment systems, and libraries. Moreover, with advancements in satellite communication, crew members can now regularly connect with their families, alleviating feelings of isolation.

Fair Wages and Benefits

Ensuring competitive wages, insurance coverage, and other benefits is central to retaining and motivating crew members. Organizations such as the International Transport Workers' Federation (ITF) play a crucial role in the negotiation of equitable terms and conditions for seafarers on a global scale.

2. Training and Skill Development: Adaptive Training Modules

With rapid technological advancements, it's vital that crew members are trained to handle new systems and equipment. Customized training modules, both online and offline, cater to these evolving needs.

Safety Drills and Emergency Response

Regular drills ensure that the crew is prepared for emergencies, be it piracy threats, fires, or medical emergencies. Such training is often mandated by international conventions like the Safety of Life at Sea (SOLAS).

Soft Skills and Leadership Training

As the industry shifts towards a more collaborative and flexible management structure, training modules now also focus on soft skills, conflict resolution, and leadership development.

3. Equal Opportunities: Gender Diversity

Historically male-dominated, the maritime sector is now actively promoting gender diversity. Efforts such as the "Women in Maritime" campaign, led by the International Maritime Organization (IMO), are designed to enhance the agency and participation of women in the maritime sector, with the objective of promoting gender equality and providing equitable opportunities for women both on board vessels and inside maritime institutions.

Non-discrimination Policies

Leading maritime companies have robust non-discrimination policies, ensuring that crew members are selected, rewarded, and promoted based on merit, irrespective of their nationality, religion, or background.

Inclusivity Initiatives

Recognizing the value of diverse perspectives, many maritime organizations now have inclusivity programs, training sessions, and awareness campaigns to ensure a harmonious and inclusive working environment.

Implications

The maritime industry's ethical treatment of its crew goes beyond mere compliance with international regulations. It's a testament to the industry's recognition of the crew's invaluable contributions and its commitment to fostering an environment that values, nurtures, and rewards every individual's dedication and hard work.

Addressing Challenges: Piracy, Illegal Trading, and Maritime Security

The tempting call of the sea, while promising bounty and adventure, also carries with it inherent challenges. The vast, often unmonitored expanses of our oceans have historically been arenas for piracy, smuggling, and other illicit activities. In the modern era, with globalization and technological advancements, the stakes are even higher, and the challenges are more complex.

Initiatives

1. Advanced Surveillance Real-time Monitoring Systems

Many ships today incorporate sophisticated radar systems, underwater sonar, and aerial drones to provide a continuous monitoring net, ensuring any threat is detected well in advance.

AI-Powered Threat Detection

Artificial intelligence algorithms, trained on historical data, can predict and detect abnormal activities or unidentified vessels approaching, providing early warning signals.

Secure Communication Channels

Encrypted satellite communication ensures that ships can communicate with their home bases and relevant authorities without the fear of interception or jamming.

2. Collaboration with International Bodies Shared Intelligence

Collaboration with international bodies like INTERPOL and the International Maritime Organization allows ships to access shared intelligence on piracy hotspots, smuggling routes, and potential threats.

Joint Naval Patrols

Many high-risk areas see joint naval patrols by multiple countries, ensuring a collective response to threats and a show of unified deterrence.

Safe Corridors

In particularly vulnerable areas, safe corridors are established where ships can travel under the protection of naval vessels, ensuring their safety from potential pirate attacks.

3. Strict Adherence to Trading Laws Transparent Cargo Manifests

By maintaining detailed and transparent cargo manifests, ships ensure that their cargo is above board and can be inspected by authorities at any point.

Regular Audits & Inspections

Regular internal audits and adherence to third-party inspections ensure that ships comply with all international trading norms and aren't inadvertently part of illicit trading networks.

Training & Awareness Programs

Crew members are trained to recognize signs of illegal goods, human trafficking, or other illicit activities, ensuring a vigilant and informed first line of defense.

Implications

The challenges of piracy, illegal trading, and maritime security aren't just operational hurdles; they're ethical imperatives. The establishment of crew safety, compliance with regulations regarding transported commodities, and the maintenance of overall security in marine operations are fundamental principles that underpin an ethical and sustainable maritime business. The marine industry is making progress toward enhancing safety and security in the world's oceans by leveraging technological breakthroughs, fostering international collaboration, and implementing rigorous regulatory measures.

The Future of Maritime Management

As we navigate deeper into the 21st century, the maritime industry is not just riding the waves but shaping them. Leveraging technology, embracing sustainability, and ensuring the well-being of its workforce, the maritime sector is poised for transformational change. Let's cast an eye toward the horizon and envision the future of maritime management.

Balancing Automation and Human Decision Making

The increasing integration of AI and automation into maritime operations presents both opportunities and challenges. While automation can streamline operations and enhance efficiency, it cannot replace the nuanced decision-making and years of experience of human operators.

Role of Automation

Advanced algorithms and sensors are automating tasks such as navigation, monitoring ship health, and even docking procedures. This not only improves efficiency but also reduces the scope for human error.

Preserving Human Touch

While automation handles routine tasks, crucial decisions, especially during emergencies, still rely on the expertise of seasoned maritime professionals. Their judgment, born out of experience, can't be replicated by machines.

Collaborative Model

The ideal future will see a collaborative model where automation handles data-heavy tasks, and humans step in for context-based decision-making, ensuring optimal outcomes.

Continuous Learning: Importance of Training and Development in the Digital Age

As maritime operations evolve, so does the need for continuous learning and training. With technology permeating every facet of operations, equipping the workforce with the latest skills becomes paramount.

Digital Training Modules

Virtual Reality (VR) and Augmented Reality (AR) based training modules can provide hands-on experience without the risks, ensuring crew members are well-prepared.

Regular Skill Upgradation

As new technologies get onboarded, regular training sessions ensure that the crew is always equipped to handle the latest tools and technologies.

Soft Skills Training

With diverse crews and a greater emphasis on collaboration, training isn't just about technical skills. Soft skills, especially communication and teamwork, become equally crucial.

Best Practices for Ethical and Efficient Maritime Management

Ethics and efficiency are no longer mutually exclusive. In fact, in the modern maritime world, they're deeply intertwined.

Guidelines: Transparency

Whether its cargo manifests, crew welfare measures, or environmental initiatives, transparency ensures trust among stakeholders and adherence to global standards.

Stakeholder Engagement

Regular dialogues with stakeholders, be it crew members, regulatory bodies, or trade partners, ensure that operations remain aligned with broader industry goals and standards.

Sustainability

From fuel-efficient operations to waste management, sustainability isn't just an ethical imperative but also a business necessity, ensuring long-term viability in a rapidly evolving global sector.

The future of maritime management will be characterized by a harmonious blend of technology and human touch, continuous upskilling, and an unwavering commitment to ethical operations. As the industry sails into this promising future, these best practices will serve as the guiding stars, ensuring a journey that's efficient, ethical, and exemplary.

Chapter 4

CREWING IN THE
NEXT DECADE

The maritime industry, paramount to global trade, is anchoring into an era where digitalization, automation, and data-driven decision-making are pivotal. Prompted by a pursuit of operational efficiency, a commitment to sustainable practices, and the complexity of risk management in maritime operations, technological innovations are no longer optional but imperative. As the sector navigates through these dynamic waves, it encounters new horizons that redefine traditional operations and systems.

Anticipated Changes in Crew Roles Amidst Technological Advances

Concurrently, the incorporation of advanced technologies such as Artificial Intelligence (AI), automation, the Internet of Things (IoT), and blockchain is reshaping the roles and skill sets required from maritime crews. The emergence of these technologies transcends mechanical and operational adjustments, extending into decision-making realms that were traditionally human. Crew roles are transitioning from manual operations to more supervisory and management-oriented responsibilities, emphasizing the need for technological proficiency, analytical thinking, and agile problem-solving in the face of real-time challenges.

In the ensuing chapter, we will dissect the multifaceted transformations occurring in the maritime sector, exploring the confluence of technological advancements with human roles and delineating strategies that ensure the harmonious evolution of maritime operations in the future.

The Transforming Role of Human Capital in Maritime Companies

From traditional sailors to technologically savvy operators In the contemporary maritime scenario, the pivotal transition from classic maritime roles to technologically nuanced operators delineates a profound metamorphosis within the industry. The traditional sailor, once primarily oriented around manual operations and a wisdom-driven understanding of the seas, is now evolving. Today's mariners are now being molded into sophisticated operators, well-versed in employing and optimizing state-of-the-art maritime technologies. This involves an amalgamation of time-honored seafaring wisdom with a profound mastery of modern-day technologies such as satellite navigation, automated systems management, and data analytics.

In practice, this transition manifests in numerous ways. For instance, modern-day sailors must understand the intricacies of global navigation satellite systems, electronic chart display and information systems (ECDIS), and automated identification systems (AIS). Their role transcends the mere operation of these systems, extending into areas such as data interpretation, which requires a nuanced understanding of various data points to make informed, real-time navigational decisions.

Managing technological assets: New responsibilities

As technology infiltrates maritime operations, the duty of crews has expanded from straightforward operation to encompass comprehensive management and preservation of technological assets. This ensures seamless operation and data integrity amidst the sprawling digital sector of contemporary maritime operations.

To illustrate, consider the maritime communication network, which has evolved beyond traditional radio communication to incorporate satellite and internet communication, creating a complex network that demands specialized knowledge for management and troubleshooting. Moreover, mariners are now tasked with ensuring the operational health and cybersecurity of complex systems that control everything from navigation to cargo management. This demands a diversified skill set that encompasses IT management, cybersecurity protocols, and data analytics, to name a few.

Addressing the cybersecurity aspect, crews are not only tasked with ensuring the physical security of the vessel but also safeguarding its digital infrastructure. This encompasses the activities of monitoring network access, ensuring data security, and deploying defensive measures to mitigate any cybersecurity risks. The global shipping sector requires understanding and compliance with global data privacy laws like the EU's General Data Privacy Regulation (GDPR).

The shipping industry's use of technology is distinguished by its complex and diverse integration. Therefore, maritime staff members must possess an enhanced skill set to ensure that operations are efficient and optimized and adhere to international standards and regulations while maintaining security. This synthesis of traditional maritime expertise with sophisticated technological acumen heralds a new era for the industry, promising enhanced capabilities, efficiency, and global operational coherence.

Strategic Management of Human Capital

Harnessing Experience and Expertise in the Age of Automation. In the crucible of the technologically-driven maritime sector, the melding of veteran expertise and novel automation becomes a strategic imperative for optimizing operations and decision-making processes. Although automation and artificial intelligence (AI) have dramatically reshaped the operational dynamics, the human element—embodied by experience, intuition, and contextual decision-making—remains irreplaceable.

The convergence of experienced professionals and automation heralds a complex operational modality where human oversight augments technological precision. Seasoned mariners, with their profound understanding of the oceanic environment, unpredictabilities, and practical seafaring knowledge, offer invaluable insights that technology alone cannot discern. Therefore, strategies to optimize this amalgamation involve not simply adopting automation but skillfully intertwining it with human expertise to inform, guide, and enhance automated processes. This interaction enables a scaffold that supports data-driven decision-making while also accommodating the adaptive, problem-solving capabilities of human operators.

The objective here extends beyond mere operational management to encompass aspects like crisis management, strategic planning, and adaptive problem-solving, where human insights, derived from years of maritime experience, guide technological applications to not just react, but proactively adapt to evolving maritime scenarios.

Employee Retention and Motivation Strategies

Effectively operating within the challenging and demanding environment of the maritime industry requires more than just acquiring a competent staff. It also entails ongoing efforts to retain and consistently motivate qualified employees. Thus, maritime entities are challenged to formulate and implement strategies that not only attract talent but also foster an environment conducive to ongoing development, satisfaction, and long-term engagement.

A fundamental aspect of this involves continuous training and development programs. Given the rapid technological advancements within the sector, ensuring that personnel are not merely conversant but proficient in utilizing and optimizing these technologies becomes critical. Tailoring training programs to address not only the technical skills but also to foster an adaptive mindset ensures that the workforce can navigate the technological advancements with agility and competence.

Competitive remuneration and benefits packages, while fundamental, need to be complemented by recognition and progression opportunities. Maritime entities must recognize and reward innovations, adaptability, and accomplishments, establishing a culture where contributions are acknowledged and celebrated. This extends into creating pathways for career progression, where ongoing development and advancement opportunities are transparent and accessible.

Moreover, considering the unique challenges of maritime professions—such as extended periods away from home and the physical and mental demands of seafaring roles—implementing and prioritizing strategies that support work-life balance and mental well-being are crucial. This could materialize through policies that manage on-board tenures, provide psychological support, and ensure periods of rest and recuperation.

In synthesizing these strategies, maritime enterprises establish a holistic, supportive environment that not only retains talent but also nurtures its ongoing development and well-being, ensuring a sustainable, proficient, and motivated workforce well into the future. This not only optimizes current operations but strategically positions companies to adeptly navigate the future maritime sector.

Building a Culture of Innovation

Inculcating a Tech-Positive MindsetBuilding an innovative culture within the maritime industry, particularly amidst rapid technological advancements, mandates an intentional and strategic shift towards embedding a tech-positive mindset across all levels of operations. Cultivating this mindset is not merely about upgrading systems and adopting new technologies but fostering an environment where technological enhancements are perceived and interacted with as catalytic enablers of operational efficacy, safety, and strategic progression.

To inculcate a genuinely tech-positive mindset within the maritime workers, it becomes imperative to focus on demystifying technology and aligning it with their daily operations and objectives. This extends to providing comprehensive training that is not only functionally relevant but also contextually applicable, ensuring crew members comprehend the tangible impacts and advantages these technological advancements introduce to their roles and broader maritime operations.

Moreover, embedding this mindset transcends practical applications to permeate perceptions, attitudes, and interactions with technology. Establishing platforms where experiences, insights, and challenges related to technological applications can be shared, discussed, and resolved fosters an environment where technology is collectively perceived as an enabler and ally, rather than an imposed disruptor.

Fostering an Environment for Continuous Improvement and Adaptation

Maritime operations are always evolving and changing; thus, an organizational culture of continual development and adaptability is needed to be relevant and effective. Such a culture not only optimizes current operations but critically positions maritime entities to adeptly navigate, adapt to, and leverage future innovations and challenges.

To cultivate this, organizations must implement robust frameworks for continuous professional development, ensuring that personnel are not only proficient with current technologies but are also equipped with the skills and knowledge to adapt to future advancements. This involves creating targeted training initiatives that not only address functional skills but also foster adaptive, forward-thinking mindsets that anticipate and navigate future innovations and challenges.

Moreover, developing a culture that champions continuous improvement and innovation involves transforming feedback and experiences into actionable insights and strategic developments. Implementing mechanisms to collect, analyze, and action feedback and experiences related to technological applications and operational processes ensures that innovations and improvements are not only iterative but also aligned with practical, on-ground realities and challenges.

Furthermore, fostering an innovative, adaptive culture also necessitates the creation of platforms and channels that facilitate and encourage knowledge sharing, collaborative problem solving, and collective innovation. Establishing forums where maritime personnel can share insights, challenges, and solutions related to technological applications, operational processes, and strategic advancements fosters an environment where collective experiences and expertise inform, enhance, and drive continuous innovation and improvement.

Through the strategic amalgamation of these initiatives, maritime entities can cultivate an organizational culture that is not only aligned with current technological and operational paradigms but is also proficiently positioned to navigate and leverage the opportunities and challenges presented by the future maritime industry.

Automation, AI, and the Future of Crewing

Specific areas of operation impacted by automation. Automation is weaving its influence across a myriad of operations within the maritime sector, bestowing enhancements that cater to both functionality and safety. The depth of its impact is varied, marking significant strides in:

Navigational Systems

Automation not only contributes to precision but also adds a layer of safety to navigation. Implementation of technologies like the Automatic Identification System (AIS) and Electronic Chart

Display and Information System (ECDIS) have enriched navigational accuracy and safety, providing real-time data and analytics that assist in informed decision-making.

Maintenance and Repair Operations

Automated drones and robots are being deployed for inspecting and performing maintenance tasks in areas that might be hazardous for human crew, minimizing risks and enhancing the predictive maintenance capabilities through continuous monitoring and data analysis.

Cargo Handling

Automated cranes, conveyors, and even autonomous vehicles within the ship ensure a more organized, timely, and efficient cargo handling, reducing turnaround times and enhancing operational efficiency. Advanced systems employ algorithms to optimize cargo placement, ensuring stability and optimal space utilization.

Communication Systems

Automation ensures steadfast and reliable communication through technologies like Maritime Satellite Technology and VHF Radio systems, which are pivotal in maintaining connections in the expansive and often treacherous sea environment.

Safety Protocols

Automated safety systems, such as fire detection and suppression systems and man-overboard detection, offer quick response times, often mitigating issues before they can escalate and ensuring the safety of the crew and vessel.

Energy Management

Smart grids and energy management systems onboard aid in optimal energy utilization, reducing operational costs and minimizing the environmental impact by efficiently managing the energy production and consumption.

Case Studies of Automation in Maritime Companies

Exploring further into real-world applications:

Yara Birkeland

The distinction of Yara Birkeland does not only lie in its autonomous capabilities but also in its adoption of an electric propulsion system. This shift towards a more ecologically balanced operational model demonstrates a foresight into an industry that leans into sustainability. Yara Birkeland is expected to replace 40,000 truck drives a year, reducing NOx and CO_2 emissions in a significant leap towards sustainable maritime logistics.

Svitzer Hermod

The Svitzer Hermod project is crucial in unfolding the tangible possibilities of remote vessel operation. In 2017, Rolls-Royce and Svitzer demonstrated successfully how a tugboat, Svitzer Hermod, can be controlled remotely. This not only marked a milestone in autonomous maritime operations but also showcased how such technology can be utilized in conducting complicated maneuvers, such as berthing and navigation through confined areas, remotely.

Hapag-Lloyd's Sajir

A noteworthy mention would be Hapag-Lloyd's Sajir, a mega-container ship that was under conversion to be powered by LNG, a cleaner alternative to the traditionally used heavy fuel oil. Sajir is enabled with a state-of-the-art ship engine control system, allowing it to be navigated from the shore, pointing towards an imminent evolution where ships could be controlled from a centralized, land-based control hub, ensuring enhanced safety and better data management.

Each case unearths various facets of automation, from autonomy to remote operations to sustainable propulsion, depicting a future where maritime operations are not just technologically advanced but are also steering towards sustainability and enhanced safety protocols.

Implications of Crew Structure and Functions

Modifying crew sizes and onboard roles the fusion of automation and Artificial Intelligence (AI) within the maritime domain is not only transforming the operational sector but is also casting a profound impact on crew structure and roles onboard.

Redefined Roles and Skillsets

The incursion of automation into various facets of maritime operations, such as navigation, maintenance, and cargo handling, calls for a redefinition of roles within the crew structure. A crew member's role metamorphoses from being a singularly skilled operator to a multi-faceted manager, adept at overseeing automated systems, analyzing data, and intervening astutely when necessary.

Specialization in Technology Management

With ships becoming conglomerates of sophisticated systems, crew members are expected to develop specialized skills in managing diverse technologies, encompassing system management, cybersecurity, and diagnostics.

Bridging the Skill Gap

This transformation isn't unilateral; maritime training institutions and on-the-job training programs now shoulder the responsibility of developing a curriculum that arms the seafarers with the requisite skills to navigate through the technological sector onboard.

Leadership in the Age of Automation

Furthermore, leadership roles onboard will require a blend of traditional maritime knowledge and astute understanding of technological applications, ensuring a balanced approach to managing both human and technical resources effectively.

Safety and Contingency Planning in Automated Operations

When discussing safety and contingencies in the context of automated operations, it's pivotal to analyze both the advantages and the potential pitfalls of this technological integration.

Enhanced Safety Through Precision

Automated systems, with their precision and consistency, mitigate risks associated with human error and enhance safety in operations. This extends from accurate navigational systems, automating dangerous maintenance tasks, to predictive analytics that can forewarn crews regarding potential issues, facilitating preemptive actions.

Dependency and System Failures

However, an over-reliance on automated systems necessitates solid contingency plans to navigate through system failures or anomalies. Although automated systems are generally considered to be highly dependable, they are not immune to potential technical malfunctions, cyber-attacks, or failures under extreme circumstances.

Preparing for Manual Override

To prioritize safety, it is imperative for crew members to possess a high level of proficiency in manual operations. This proficiency is necessary to guarantee uninterrupted operations and safeguard both the vessel and her crew in the case of a malfunction in the automated system. This involves regular drills, training, and scenarios where they manually navigate through operations, ensuring that the skills do not attenuate through disuse.

Ethical and Practical Considerations

Moreover, practical and ethical considerations such as decision-making in critical situations, emergency response, and ensuring the well-being of the crew must be meticulously planned, balancing technological inputs and human judgment harmoniously.

Regulatory and Compliance Aspect

The shifting scenario also brings with it a plethora of regulatory considerations, ensuring that automated operations comply with international maritime laws and standards ensuring that safety, security, and environmental responsibilities are judiciously adhered to.

In this balance of technology and human interface, crafting a stable operational model, where automated systems and human oversight coalesce seamlessly, will sculpt the future of safe, efficient, and sustainable maritime operations.

Balancing Automation and Human Expertise

Let us have a look at how to seamlessly integrate and collaborate between man and machine

The Harmonization of Capacities

In the confluence of automation and human expertise, establishing a harmonious operation is paramount. It's imperative to intertwine human decision-making capabilities with the precision and consistency of automated systems, crafting an operational paradigm where each complements the other.

Transparent Automated Processes

Integrating interfaces that facilitate intuitive understanding and oversight of automated processes for crew members is vital. This ensures that even complex automated processes are transparent and can be monitored and controlled effectively by the crew.

The system should offer comprehensive visibility into operational data, ensuring that crew members can make informed decisions by gauging the performance and outputs of automated systems.

Collaborative Decision-Making

Incorporating mechanisms for collaborative decision-making, where automated systems provide data-driven inputs and human operators contribute experiential and strategic insights, ensures that decisions are both accurate and judicious.

Establishing protocols where human oversight can seamlessly intervene and guide automated systems in scenarios that are complex, ambiguous, or fall outside the pre-defined parameters of the automated system.

Strategic Interventions

Training programs and operational protocols must be established to ensure that human operators can strategically intervene, modifying or overruling automated processes when requisite and ensuring the optimal alignment with broader operational goals and safety considerations.

Ethical Considerations in Automation: Job Displacement and Upskilling

Navigating through the ethical maelstrom that automation in the maritime sector brings forth, particularly revolving around job displacement and workforce management, necessitates a morally cognizant and strategically viable approach.

Ensuring that the transition to more automated operations doesn't inadvertently marginalize or displace existing crew members demands meticulous planning and ethical foresight.

Phased Introduction of Automation

Implementing a phased introduction of automated technologies that allows adequate time for existing crew to understand, adapt, and upskill to new operational paradigms without feeling abruptly displaced. Engaging in thorough communications with the crew about the reasons, benefits, and strategies for introducing automation, ensuring transparency and avoiding unwarranted anxiety among the workforce.

Transparent Communication and Involvement

Fostering an environment of transparency and inclusivity where the transition to automation is communicated clearly, and feedback from the crew is earnestly considered in strategizing the transition. Inviting crew members to be active participants in the transition process, thereby acknowledging their valuable insights and enabling them to be advocates for change rather than mere spectators.

Alignment with Organizational Goals

Ensuring that the integration of automation aligns with the larger organizational objectives, which should not only prioritize operational efficiency but also safeguard the welfare and career progression of the crew.

Formulating strategies where the benefits of automation, such as enhanced operational efficiency and safety, are translated into tangible benefits for the crew, such as reduced work hours, enhanced work conditions, or better remuneration.

Upskilling as an Ethical and Operational Imperative

Creating a lattice for career progression and skill enhancement amidst an automated sector is pivotal in ensuring that the workforce evolves alongside the technology.

Tailored Upskilling Programs

Establishing upskilling and training programs that are tailored to the existing skillsets and future needs of the crew, ensuring that they are relevant, accessible, and practically beneficial in the new technological environment.

Investing in continuous learning platforms that not only address the immediate technological transition but also inculcate a culture of perpetual skill enhancement among the crew.

Career Progression Pathways

Developing clear and viable career progression pathways for crew members in the automated paradigm, ensuring that their experience and expertise are leveraged and acknowledged in the new operational context.

Formulating strategies that enable crew members to transition into roles where their maritime experience enhances the effectiveness and strategic deployment of automated systems.

Ensuring Equity and Inclusivity in the Transition:

The transition to an automated future must be navigated with a steadfast commitment to ensuring equity and inclusivity.

Equitable Access to Opportunities

Crafting policies and practices that ensure all crew members, irrespective of their existing roles, have equitable access to upskilling and career progression opportunities in the automated future.

Instituting mechanisms to ensure that biases, whether implicit or explicit, do not mar the accessibility or effectiveness of upskilling and transition opportunities.

Inclusivity in Technological Advancement

Emphasizing the creation of an inclusive environment wherein technological advancements are viewed and experienced as collective progress rather than isolative or exclusionary progression.

Establishing support systems that cater to diverse learning paces and styles, ensuring that all crew members, regardless of their starting point, can find a path to advancement and adaptation in the automated future.

In essence, the path towards automation must be tread with an unwavering commitment to the ethical treatment, upskilling, and career progression of the human workforce, ensuring that the technological advancements of the maritime sector are firmly anchored in principles of fairness, inclusivity, and mutual progression. This not only safeguards the welfare of the workforce but also ensures that the experiential wisdom of seasoned mariners continues to steer the industry, albeit in a new, technologically-augmented context.

Training & Education: Preparing Crews for the Future

In the waves of modernity, as the maritime sector engulfs more technology and automation, the essence and structure of training and education must be redefined to maintain relevance and efficacy. Crafting a future-ready crew demands the integration of forward-looking skills and innovative training methodologies that resonate with the technological trends and challenges emerging on the maritime horizon.

Identifying the Skills of the Future

Navigating through the digitalized seas of the future demands the identification and cultivation of skills that synergize traditional maritime expertise with technological prowess.

Technical Skills: Operating and Troubleshooting Automated Systems

- The interconnectivity of maritime operations and technology dictates a strong requirement for skills that amalgamate technical understanding with practical application, such as operating, managing, and troubleshooting various automated and AI-driven systems on board.
- Facilitating a detailed understanding of system architectures, data management, cybersecurity, and mechanical interfacing to equip crews with the capabilities to oversee automated operations and intervene effectively during discrepancies or malfunctions.

Adaptive Skills: Managing Change and Uncertainty

- In a technologically fluid environment, adaptive skills – encompassing agility, problem-solving, and decision-making amidst change and uncertainty – become paramount.
- Focusing on scenario-based training that exposes crews to various technological and operational anomalies, thereby honing their skills to navigate through uncharted and unpredicted situations without compromising safety and operational integrity.

Evolving Training Methodologies

The digital age brings forth avenues to reimagine and revitalize training methodologies, incorporating technology not just as the subject matter of training but also as the medium through which it is delivered.

Leveraging Digital Platforms for Remote and Continuous Learning

- The integration of digital platforms to facilitate remote learning, ensuring that crews can access relevant knowledge and training modules irrespective of their geographical locales.
- Crafting a blend of synchronous and asynchronous learning experiences that allow crew members to balance operational duties with continuous learning endeavors, ensuring sustained skill development.

VR and AR in Practical Skills Training and Crisis Simulations

- Implementing Virtual Reality (VR) and Augmented Reality (AR) to transcend traditional training barriers, providing immersive and realistic training scenarios that simulate real-world crises and operational contexts in a safe and controlled environment.
- Utilizing VR and AR for conducting practical skill assessments and collaborative training exercises, ensuring that the skills imparted through digital platforms are contextually relevant and practically applicable on board.

Fostering a Continuous Learning Environment

Constructing a future-proof maritime workforce mandates the establishment of a learning environment that is not episodic but continuous, ensuring that the crew evolves in tandem with technological advancements.

Incorporating Learning as a Part of the Organizational Culture

- Weaving learning into the very fabric of organizational culture, where continuous skill development is viewed not as an optional endeavor but a fundamental expectation and shared commitment.
- Establishing mechanisms that celebrate and recognize continuous learning and skill development, embedding them as key performance indicators in career progression pathways.

Building Feedback Loops: Continuous Improvement of Training Programs

- Designing feedback loops where experiences and insights from the crew actively shape the evolution of training programs, ensuring they remain relevant, engaging, and operationally pertinent.
- Utilizing data analytics to gauge the efficacy and impact of training programs, identifying areas of improvement, and dynamically adapting content and methodologies to cater to the emerging needs and challenges faced by the crews.

As the maritime domain boards on an automated and technologically enriched future, the potency and relevance of training and education will be determined by its ability to foresee, adapt, and inculcate the skills and methodologies that navigate through the technological currents, ensuring that the human element of maritime operations continues to steer confidently through the digital waves.

Legal and Ethical Considerations

The journey toward automating maritime operations and reshaping crew structures intertwines with multifaceted legal and ethical considerations. This not only stands crucial for organizational integrity but also casts a significant impact on the broader socio-economic and professional maritime sector.

Navigating Legal Waters of Crewing and Automation

As maritime operations sail into more automated territories, the legal waters become challanging, demanding comprehensive navigation through international and local regulations that govern crewing and the implementation of automated technologies.

International Laws and Regulations on Crew Size and Qualifications

- The International Maritime Organization (IMO) and various regional authorities impose standards regarding crew sizes, qualifications, and working conditions, established in conventions like the Maritime Labor Convention (MLC) and Standards of Training, Certification, and Watchkeeping (STCW) Convention.
- Balancing the integration of automated technologies while ensuring compliance with mandates concerning crew qualifications, safety training, and work-rest hours stands pivotal to lawful and effective maritime operations.

Ensuring Legal Compliance in Varied Jurisdictions

- Maritime operations often span across varied jurisdictions, each potentially embedding its own legal stipulations concerning automation, crewing, and safety.
- Crafting a comprehensive legal compliance strategy that respects and adheres to the regulatory contexts of all navigated jurisdictions, employing a possibly dynamic crewing model that adjusts to the varying legal environments.

Ethical Crew Management in an Automated Future

As automation assumes a more dominant role, ethical considerations in crew management come to the forefront, affecting both the existing workforce and shaping future employment trends within the industry.

Addressing the Challenge of Job Losses Due to Automation

- Automation, while enhancing operational efficiency, harbors the potential to diminish certain traditional roles within maritime crews, posing ethical and socio-economic challenges concerning job displacements.
- Initiating dialogues with maritime unions and workers' representatives to ensure transparent communication and collaborative strategizing to mitigate the impact of automation on employment within the sector.

Creating Ethical Transition Plans for Existing Crews

- Developing and implementing transition plans for existing crews that might be impacted by the integration of automated technologies, ensuring their skills and expertise continue to find relevance within the evolved operational models.
- Facilitating upskilling and reskilling programs that enable existing crew members to adapt to the emerging technological paradigms, ensuring that their career trajectories within the maritime domain remain robust and upward.

The legal and ethical aspects of crewing in an automated future require a coordinated strategy that follows regulatory mandates and protects and nurtures the human element in marine operations. Maritime businesses may combine technical innovation with legal and moral integrity by integrating strict legal compliance with ethical transition and management methods. This multidimensional approach protects business reputation and builds a stable, skilled, and ethical workforce to navigate future technology.

Case Studies

Examining real-world case studies illuminates practical insights, delivering a realistic glimpse into the obstacles and achievements encountered by entities on this transformative journey in understanding the impact and navigational path of AI and automation in marine contexts.

Implementing AI and Automation: Challenges and Triumphs

In-depth investigation of the complexities of technology deployment can yield multiple learnings, diving into both successes and obstacles encountered.

Yara Birkeland

- A pioneering endeavor, the Yara Birkeland project showcases a notable stride towards fully autonomous ship operations, presenting a model wherein zero-emission, automated freight transport becomes plausible.
- Challenges navigated encompass technological hurdles, safety assurance, and regulatory compliance, forging a path that marries technological innovation with lawful and safe operations.

Maersk and IBM's TradeLens

- Maersk and IBM embarked on a collaborative venture, TradeLens, leveraging blockchain technology to enhance transparency and efficiency within maritime supply chains.
- While fostering enhanced data visibility and operational efficiency, navigating through the integration challenges and ensuring broad industry adoption presented complicated hurdles.

Adapting to Change: Stories of Successful Transition

Examining stories of successful transition and adaptation to technological enhancements reveals not only the strategic undertakings but also the human and operational impacts of such shifts.

Carnival Cruise Line's Ocean Medallion

Implementing a wearable IoT device, the Ocean Medallion, Carnival enhanced customer experiences while presenting a change management model wherein staff adapted to a technologically augmented operational framework.

The transition necessitated comprehensive staff training and a reorientation towards a tech-augmented service model, offering insights into managing change amidst technological integration in maritime contexts.

Svitzer's Remote Navigation Triumph

- Svitzer's successful remote navigation of a vessel through a harbor points towards the potential within remote and automated navigational technologies.
- Adapting crew roles and ensuring safe and effective remote operations necessitated not only technological but also operational and personnel-focused strategies, illuminating the multi-dimensional aspects of such transitions.

Concluding Thoughts

Sailing across the infinite sea of revelations, lessons, and useful case studies, the maritime industry sets sail for an automated future, firmly grounded by a number of critical factors that come from an inventive and complex excursion. Technology brings human skills together to maintain balance in automated operations, protect vital job sectors, and navigate tricky regulatory and ethical waters. The future is bright and full of opportunities for technological advancement, operational efficiency, and a completely new paradigm for maritime operations. Therefore, it is imperative that human capital remain at the center, skillfully steering the industry with a combination of knowledge, moral judgment, and unmatched strategic vision. As the maritime industry goes on this promising yet advanced sail, it stands on the brink of not only redefining its operational models but also refining its professional and ethical foundations, charting a course toward a future that promises to illuminate technological brilliance, guided by human expertise and ethical veracity.

Chapter 5

THE TECHNOLOGICAL
REVOLUTION IN MARITIME

The maritime sector, being a vital cog in the global trade machine, has always been influenced significantly by technological advances. However, in the context of the digital era, where data and connectivity have redefined traditional operational matrices, digital transformation becomes indispensable. Factors, such as streamlined operations, data-driven decision-making, enhanced safety protocols, and customer satisfaction, delineate its criticality. Digital transformation in the maritime sector is not just an operational upgrade but a strategic necessity, essential to maintaining competitiveness, ensuring safety, and steering through the regulatory and environmental demands which have become more stringent in recent years. From utilizing data analytics for optimized route planning, reducing fuel consumption and emissions, to implementing blockchain for transparent and secure documentation processes, digital initiatives have become imperative for operational efficiency and regulatory compliance.

Evolution from Traditional Practices

The shift from traditional to digitalized operations in maritime encapsulates a holistic transformation that transcends mere technological adoption. It is a transformation that alters the structural, cultural, and operational dimensions of maritime organizations. Traditional practices, characterized by manual processes, decentralized data, and a reliance on experiential decision-making, have been underscored by inherent challenges such as operational inefficiencies, increased error margins, and vulnerability to market volatilities. Digital transformation, in contrast, brings forth centralized data management, predictive analytics, automated operations, and real-time monitoring, enhancing operational visibility, accuracy, and efficiency.

In this evolution, a transition is observed from isolated operational silos to interconnected, data-driven models. For instance, advancements like the adoption of cloud computing have facilitated centralized data storage and access, ensuring that decision-makers can derive insights from a unified data pool. Similarly, the integration of IoT devices enables real-time monitoring of ship health, cargo status, and navigational parameters, which were traditionally managed through manual logs and periodic checks. Automated systems, enabled by AI, have begun to take over routine tasks, minimizing human error and freeing up the crew for strategic and oversight roles. The evolution also entails a shift towards a more proactive approach to risk management, where predictive analytics and automated alerts enable pre-emptive measures against potential operational and safety risks.

This journey from traditional to digital practices in the maritime sector doesn't only embody technological alterations but also demands a reconfiguration of organizational culture, workflows, and skillsets, ultimately heralding an era where technology and human expertise sail collaboratively towards operational excellence and innovative horizons.

This comprehensive transition implies a necessary turmoil of legacy systems and a reevaluation of operational norms aimed at aligning the maritime sector with the digital future, ensuring it keeps pace with technological advancements and leverages them to navigate the challenges and opportunities that lie ahead in the vast ocean of global commerce.

Real-world Impacts of Digital Transitions

The migration towards digital outlooks in the maritime industry has generated a range of palpable and transformative impacts across various facets of operations, strategy, and stakeholder interactions. Digital transformations have significantly impacted marine enterprises by enhancing operational efficiencies, implementing robust risk management practices, optimizing logistical processes, and ultimately improving customer experiences. These transformations have closely interwoven the operational and strategic aspects of these firms. Success Stories and Lessons Learned from Digital Transformations

Mediterranean Shipping Company (MSC)

MSC, one of the world's largest shipping companies, embarked on a digital transformation journey that centered on enhancing operational efficiencies and customer experiences through technology. MSC has implemented a digital platform that leverages IoT and blockchain to enhance transparency and traceability in its supply chains. MSC's implementation of TradeLens, a blockchain-enabled digital shipping platform developed in collaboration with IBM and Maersk, revolutionized its cargo tracking, documentation processes, and stakeholder communication, ensuring secure, immutable, and transparent data sharing.

Lessons Learned

Here are the lessons that we can learn from these pioneer shipping companies.

Inter-organizational Collaboration

Working with industry peers and technology experts can yield mutually beneficial technological advancements.

Customer-Centric Innovation

Technology should be leveraged not just for internal efficiencies but also to enhance external stakeholder interactions and experiences.

Security in Transparency

While digital platforms can enhance transparency, ensuring data security and privacy is paramount to maintain stakeholder trust.

Maersk

Navigating through Cyber Challenges

Maersk, a global leader in container logistics, endured a massive cyberattack in 2017, which significantly impacted its operations. The NotPetya malware disrupted the company's IT systems worldwide, affecting terminals, logistics, and ancillary services.

Digital Resilience and Recovery

Despite the turmoil, Maersk managed a remarkable recovery and simultaneously exhibited transparency, consistently communicating with stakeholders throughout the crisis. Post-recovery, Maersk prioritized bolstering its cybersecurity infrastructure, investing in systems and practices that safeguarded them against future cyber threats.

Lessons Learned

Here are the lessons that we can learn from these pioneers shipping companies.

Preparedness for Cyber Threats

The pivotal need to invest in cybersecurity infrastructure to guard against the potential devastation of cyberattacks.

Transparency in Crisis

Maintaining clear, transparent communication with stakeholders during crises can preserve trust and reputation.

Learning and Evolving

Using setbacks as springboards to enhance and fortify operational and security protocols, ensuring learning and evolution in response to challenges.

Through these case studies, the maritime industry witnesses the duality of digital transformation, where MSC exemplifies the enhancement of operational and customer interfaces through technology, while Maersk provides crucial insights into navigating through and evolving from digital crises. Both stories, albeit varied, underline the pivotal roles of preparation, transparency, security, and continual evolution in the digital transformation journey, highlighting not just the opportunities but also the intricate challenges navigated by entities sailing through digital seas.

Predicting Future Trends and Innovations

In the maritime industry's digital future, several trends and innovations are likely to steer the course. The adoption of Artificial Intelligence (AI) and Machine Learning (ML) is projected to significantly augment predictive maintenance, intelligent routing, and operational decision-making. Furthermore, the incorporation of the Internet of Things (IoT) will amplify data-driven operations, offering unprecedented visibility and control over various maritime aspects.

Blockchain technology is another contender, with potential applications in ensuring transparency, security, and efficiency in supply chain management. Digital twinning, wherein a virtual model of a physical entity or system is created, is set to gain momentum, enhancing monitoring, analysis, and control of ship systems. The increased use of autonomous vehicles and drones for tasks like inspections, surveillance, and possibly even short-range transport, are on the horizon as well.

The emergence of smart ports, employing digital technologies for optimized logistics, reduced environmental impact, and improved safety and security, will be vital in interlinking the digital transformation of various maritime stakeholders.

Strategies for Adapting to Ongoing Changes

Following strategies can be adapted:

Embracing a Culture of Continuous Learning

- Creating a workforce that is adaptive and continually upskilling to stay abreast of technological advancements.
- Integrating regular training and development programs that focus on emerging technologies and methodologies.

Investment in Research and Development

- Allocating resources towards R&D to explore and exploit emerging technologies and practices.
- Building collaborations with technology developers, industry experts, and academic institutions to foster innovation.

Building Robust Cybersecurity Frameworks

- Recognizing the indispensable role of cybersecurity in safeguarding operations, data, and stakeholder interactions.
- Implementing robust cybersecurity protocols and regular audits to safeguard against potential threats and vulnerabilities.

Developing Agile Operation Models

- Creating operation models that can easily adapt to technological changes and integrate new tools and practices.
- Implementing agile project management and operational methodologies that accommodate change and facilitate rapid decision-making.

Enhancing Stakeholder Collaboration and Communication

- Establishing platforms and practices that ensure clear, transparent, and secure communication amongst various stakeholders.
- Building collaborative ecosystems that enable knowledge sharing, mutual growth, and co-creation of value amongst partners, suppliers, and customers.

Legal and Ethical Preparedness

- Staying informed and compliant with international laws and regulations pertaining to digital operations, data privacy, and cybersecurity.
- Ensuring that the ethical implications of digital transformations, especially in aspects like employment and data handling, are addressed and managed with transparency and integrity.

In the foreseeable future, the maritime industry, bathed in the digital tide, will need to embody an ethos of continual adaptation, innovation, and ethical operation, ensuring that the sails are set not just towards operational excellence but also towards sustainable and responsible digital evolution.

Application in Navigation, Communication, and Operation Navigation

- **Precision and Accessibility:** IoT enables precise and real-time tracking of vessels, leveraging various sensors and GPS technologies to enhance navigational accuracy and decision-making.
- **Automated Navigation Systems:** Utilizing data from sensors, IoT-integrated systems can optimize routes by considering various factors like weather conditions, water currents, and obstacle detection, thus augmenting navigational safety and efficiency.

Communication:

- **Enhanced Connectivity:** The implementation of IoT ensures constant and robust communication between ships, ports, and stakeholders, even in the remote expanses of the oceans, leveraging satellite and wireless communication technologies.
- **Data Sharing:** Real-time sharing of vital data among ship crews, fleet managers, and other pertinent entities enhances coordinated operations and decision-making.

Operation:

- **Predictive Maintenance:** IoT devices can monitor machinery and system health, predict potential failures, and thereby optimize maintenance schedules and prevent unexpected downtimes.
- **Energy Management:** IoT aids in monitoring and managing energy usage on vessels, ensuring optimal fuel consumption and reducing environmental impact.
- **Cargo Management:** IoT applications monitor the status and conditions (such as temperature and humidity) of cargoes, ensuring that they are maintained in optimal conditions and alerting to any deviations.

Case Studies Illustrating Effective IoT Utilization

Maersk Line

- To ensure optimal operational and energy efficiency, Maersk uses Internet of Things (IoT) technologies to monitor its fleet and containers globally.
- Smart containers embedded with IoT devices allow Maersk to ensure the quality of perishable goods through real-time monitoring and management of container conditions.
- IoT sensors track engine performance, weather conditions, and cargo status, among other variables, allowing Maersk to optimize routes and maintain cargo integrity.

MSC (Mediterranean Shipping Company)

- To improve container cargo visibility and management, MSC has deployed smart container systems that make use of the Internet of Things. The temperature, humidity, and location of the goods—especially perishable or sensitive ones—are monitored by sensors built into the smart containers to guarantee that they are transported under ideal circumstances.
- The utilization of real-time data obtained from IoT devices enables MSC to promptly implement corrective measures in the event of deviations, thus guaranteeing the safety of cargo and enhancing customer satisfaction.

Rolls-Royce and the Autonomous Ship Development

- Although focused on AI, Rolls-Royce's venture into autonomous ships is deeply intertwined with IoT. The sensors and connectivity that IoT provides form the foundation upon which autonomous ships perceive their environment and make intelligent decisions.

- This involves not only internal operations and navigation but also communication with other vessels and port facilities, ensuring synchronized and safe operations.

The structured integration of IoT in navigation, communication, and various operational aspects elucidates a future where technology and human expertise sail together towards a horizon of innovation and enhanced maritime capabilities.

Implementing AI in the Maritime Spectrum

Artificial Intelligence (AI) in the maritime industry heralds a sea change in how operations, decision-making, and automation are approached, converging into an ecosystem where technological precision and human oversight navigate in tandem.

Decision-making

- **Predictive Analytics:** AI algorithms analyze vast datasets to predict outcomes and trends, such as optimal routes, delivery times, and potential maintenance issues, thereby enhancing planning and decision-making.
- **Risk Management:** Through analysis of historical and real-time data, AI enables identification and mitigation of risks, offering solutions or alternatives during perilous situations or unforeseen challenges.
- **Operational Optimization:** By leveraging AI, maritime operations can be optimized through intelligent scheduling, route planning, and resource allocation, ensuring efficiency and cost-effectiveness.

Automated Operations:

- **Autonomous Vessels:** AI facilitates the development and operation of autonomous ships, capable of self-navigating across oceans with minimal human intervention, governed by algorithms capable of making real-time navigational decisions.
- **Automated Cargo Handling:** AI-driven systems can automate cargo handling, including loading, unloading, and inventory management, ensuring accuracy and reducing manual labor.
- **Energy Management:** AI systems can autonomously manage and optimize energy usage, ensuring eco-friendly and cost-effective operations.

Examining Case Studies and Analyzing Outcomes

Yara Birkeland

- The Yara Birkeland, often hailed as the world's first fully electric and autonomous container ship, marks a landmark case of AI application in the maritime sector. Designed to transport fertilizers from Yara's plant to nearby ports in Norway, the vessel is envisaged to replace 40,000 truck journeys annually, dramatically reducing NOx and CO2 emissions.
- Key Outcomes: While still in development, the project symbolizes a step towards sustainable shipping, marrying autonomous operations with electric propulsion, thereby potentially reducing operational costs and minimizing environmental impact.

Rolls-Royce and Svitzer

- Rolls-Royce, in partnership with Svitzer, has demonstrated the capability of AI through their development and testing of a remotely operated commercial vessel. The Svitzer Hermod, a tugboat, was successfully operated by a captain stationed at a remote base, utilizing an array of technological systems.
- Key Outcomes: This instance highlighted the real-world applicability and potential scalability of remote and autonomous maritime operations, raising pertinent discussions about the future navigational model of the maritime industry and paving the path for further innovations and adoptions in the sector.

Maersk and IBM with TradeLens

- Maersk and IBM launched TradeLens, a blockchain and AI-powered platform designed to promote more efficient and secure global trade, bringing together various stakeholders like shippers, freight forwarders, port and terminal operators, and governments.

- **Key Outcomes:** TradeLens provides a secure and transparent platform to transform manual and paper-based documents, enabling smarter, more secure, and more efficient global trade. The use of AI enables predictive analytics, assisting stakeholders in making data-driven decisions.

It's clear that AI stands as a formidable force, navigating the industry towards enhanced operational efficacies, strategic decision-making, and sustainable practices, all the while directing through the complexities and challenges embedded in the integration of new technologies into established operational frameworks.

Relationship Between IoT and AI

The integration of Artificial Intelligence (AI) and the Internet of Things (IoT) creates a mutually beneficial partnership that drives maritime operations towards improved connection, thoughtful decision-making, and efficient operations in the future.

Enhanced Data Utilization

- **Data Acquisition through IoT:** Sensors and devices connected via IoT networks collect a vast array of data from multiple points throughout the vessel, including engine performance, navigation data, weather conditions, and cargo status.
- **Intelligent Analysis with AI:** This voluminous data is then fed into AI algorithms, where it is processed and analyzed to derive actionable insights, thus ensuring informed decision-making and predictive analytics in various operational aspects.

Predictive Maintenance

- **Monitoring Health of Systems:** IoT devices consistently monitor the health of various onboard systems, detecting anomalies or wear and tear on mechanical components.
- **Predictive Interventions:** AI algorithms analyze this data to predict potential breakdowns or required maintenance activities, facilitating pre-emptive interventions and minimizing unplanned downtimes.

Autonomous Operations

- **Real-time Navigation and Adjustments:** IoT ensures real-time data from various sensors, like GPS and radar, while AI interprets this data to make navigational decisions, optimizing routes, and avoiding obstacles or treacherous weather conditions.
- **Automated Cargo Management:** IoT devices monitor cargo conditions and manage inventory, while AI ensures optimal storage, timely loading/unloading, and adherence to compliance standards.

Future Potential and Predicted Developments

Advanced Autonomous Vessels

- **Remote Operations:** We might observe a rise in remotely operated vessels, where IoT devices provide real-time data to shore-based control centers, and AI aids human operators in managing vessels from afar.
- **Fully Autonomous Shipping:** Continuous advancements might pave the way for fully autonomous ships, navigating across the globe with minimal human intervention, powered by AI's decision-making and IoT's connectivity.

Intelligent Port Operations

- **Smart Ports:** With IoT and AI, ports can evolve into smart ecosystems where every entity, from cranes to carriers, is interconnected, sharing real-time data, and operating based on intelligent decisions.
- **Streamlined Logistics:** The duo can optimize logistics, ensuring precise timing for cargo loading/unloading, reducing dwell times, and enhancing overall port throughput.

Enhanced Safety and Security

- **Proactive Safety Protocols:** IoT and AI will work in synergy to predict and mitigate potential safety incidents, ensuring the wellbeing of crew, vessel, and cargo.

- **Cybersecurity:** With increased connectivity comes the need for robust cybersecurity. Future developments might see AI identifying and neutralizing cybersecurity threats while ensuring the integrity and security of data and onboard systems.

The future maritime sector, interconnected with IoT and AI, will glide on data-driven operations, autonomous functions, and intelligent decision-making. As the sector gradually unfolds its sails and capitalizes on technology improvements, it will chart a path toward a future that is technologically strong, operationally exceptional, and sustainably forward-thinking.

Emerging Technologies and Their Potential Impact

In technical progress, the marine industry is benefiting from various cutting-edge breakthroughs. These include above mentioned the Internet of Things (IoT) and Artificial Intelligence (AI), as well as other new technologies like Drones, Robotics, and Blockchain. Each of these innovations has the potential to significantly revolutionize different operational aspects within the maritime sector.

Drones in Maritime

Surveillance and Monitoring

- Employing drones for aerial surveillance of vessel structures and surroundings, enhancing safety and security by providing real-time visuals and data.
- Utilization in anti-piracy measures, where drones can provide a visual feed of surrounding areas, detecting potential threats at an early stage.

Cargo and Delivery Operations

- Drones can facilitate timely and efficient delivery of smaller cargo or essential supplies to vessels, particularly in instances requiring rapid deployment such as medical emergencies or critical spare part delivery.

Inspection and Maintenance

- Utilizing drones for hard-to-reach areas, conducting inspections and identifying maintenance needs without risking crew safety.

Robotics in Maritime

Automated Cargo Handling

Deploying robotics in the loading, unloading, and handling of cargo, ensuring precision, reducing manual labor, and mitigating related risks.

Underwater Operations

Submersible robots can undertake underwater inspections and repairs, alleviating human divers from the perils of deep-sea operations.

Routine Operations and Maintenance

Employing robots for routine cleaning, maintenance, and other operational tasks, ensuring consistency and freeing human crews for more strategic undertakings.

Blockchain in Maritime

Secure and Transparent Transactions Employing blockchain to establish a secure, decentralized, and tamper-proof ledger for transactions, ensuring transparency and trust among stakeholders.

Supply Chain and Logistics Management

Leveraging blockchain for real-time tracking of cargo and shipments, authenticating and documenting every transition phase from manufacturer to end-user, enhancing traceability, and accountability in the logistics chain.

Smart Contracts

Implementing smart contracts that automatically execute contractual clauses upon meeting predefined criteria, reducing bureaucratic delays, and ensuring swift, unbiased transaction executions.

Exploring Applications and Possibilities

The integration of these technologies unfolds a numerous of applications and possibilities:

Integrated Operations Centre

Creating a unified operational hub which amalgamates data from drones, robots, and blockchain, ensuring a centralized, real-time overview and management of various maritime operations.

Risk Mitigation

Employing a combination of drones for visual assessment, robots for physical interventions, and blockchain for ensuring operational and transactional integrity, all working cohesively to mitigate risks and enhance operational robustness.

Sustainable and Efficient Operations

Leveraging these technologies to reduce manual interventions, optimize operational pathways, and ensure sustainability through enhanced efficiency and reduced resource utilization.

Sailing ahead, the sector uplifts its operational capabilities and anchors itself in a future shaped by technological adeptness, strategic foresight, and sustainable endeavors, charting a revolutionary course with operational excellence and innovative proficiency.

Assessing Real-world Implementations

Implementing emerging technologies in the maritime sector carries its own set of challenges and opportunities, as demonstrated by the following cases:

Port of Rotterdam and IBM's IoT Implementation:

The Port of Rotterdam, Europe's largest seaport, partnered with IBM to implement IoT technologies, aiming to transition into a "smart port." The project utilized sensors and AI to gather data on water and weather conditions, assisting ship captains in making informed navigation decisions and consequently optimizing port schedules and functionality.

Challenges

Implementing a widespread sensor network, ensuring data accuracy, and building a robust analytics platform.

Opportunities

Enhanced operational efficiency, improved navigational safety, and boosted sustainability through optimal vessel routing and port management.

DP World's Robotics and Blockchain Adoption:

Global logistics leader DP World integrated robotics and blockchain to streamline container logistics and secure transactions across its various ports.

Challenges

Integrating technologies across diverse global operations, ensuring stakeholder buy-in, and maintaining technology reliability.

Opportunities

Increased transactional transparency, reduced errors in logistics handling, and enhanced security in operations and transactions.

Challenges and Opportunities Explored

Challenges

Integration Issues: Ensuring seamless integration of technologies into existing systems.

Data Security: Protecting the data generated and processed by technologies like IoT, blockchain, and AI.

Staff Training: Ensuring staff are proficient and adaptable to technologically-driven operational models.

Opportunities

Operational Efficiency: Streamlined processes, reduced turnaround times, and minimized manual interventions.

Enhanced Security: Providing a more secure and transparent platform for transactions and data storage.

Global Collaboration: Facilitating more cohesive and transparent global logistics and operational collaborations through a unified technological framework.

Prospective Technological Developments

The journey forward envisages the exploration and adoption of an array of technologies, such as Quantum Computing, which promises unparalleled computational prowess, and Augmented Reality (AR), offering enriched interactive experiences and enhanced operational capabilities in navigation, maintenance, and training. Furthermore, integrating technologies like 5G will provide the requisite backbone for robust, high-speed data transmission, particularly crucial for technologies like IoT and autonomous vessel operations.

Strategic Alignment with Future Technologies
Building Technological Agility

- Crafting strategies that enable the maritime sector to stay abreast of technological advancements, ensuring systematic reviews and incorporation of relevant technologies to stay ahead in the operational curve.
- This includes robust R&D, partnerships with tech firms, and investments in startups innovating in the maritime tech space.

Ensuring Ethical and Sustainable Implementations

- Incorporating technologies in a manner that not only uplifts operational metrics but also aligns with ethical considerations and sustainability objectives.
- This encompasses ensuring fair employment practices, minimizing environmental impacts, and contributing positively to global sustainability goals.

Policy and Regulatory Compliance

- Engaging in active dialogues with regulatory bodies to formulate policies that foster technological adoption while safeguarding legal, ethical, and safety parameters.
- Advocating for a globally unified regulatory framework to ensure consistent adherence and facilitate smoother international operations.

Maritime propels itself into a new era of operational excellence and ensures that its voyage is etched with the markers of innovation, integrity, and inclusive growth.

Technology in Maritime Risk Management and Safety

Enhanced Data Analysis

In maritime operations, data acts as a pivotal fulcrum, and the advent of technology has dramatically elevated the capacity for in-depth data analysis. Various tools, particularly Machine Learning and Artificial Intelligence, facilitate meticulous analysis of diverse data, such as weather conditions, vessel diagnostics, cargo status, and crew wellbeing, enabling predictive modeling and proactive decision-making. Enhanced data analysis aids in shaping informed, data-driven decisions, reducing downtimes, and averting potential crises by identifying issues even before they escalate.

Improving Predictive and Preventive Measures

Adopting a predictive approach is vital in averting potential perils in maritime operations. Technology enables continuous monitoring and predictive maintenance of vessel components, thereby reducing unexpected breakdowns and enhancing the vessel's lifespan. GPS and AIS (Automatic Identification Systems) technologies, alongside meteorological tools, empower vessels to forecast and navigate away from potential threats like rough seas, storms, or geo-political unrest areas, thereby ensuring smooth, secure, and punctual voyages.

Utilizing Technology for Safety Protocols and Emergency Responses

Technology becomes a critical ally in ensuring maritime safety. Automated systems can quickly execute emergency protocols, significantly minimizing human error. For instance, should a fire break out, automated extinguishing systems can be instantly deployed, while AI-driven mechanisms can manage evacuation protocols efficiently. Similarly, IoT devices can continuously monitor various parameters, ensuring that any irregularity or deviation activates an immediate response, thereby safeguarding both crew and cargo.

Examining Real-world Applications and Outcomes

Real-world applications of technology in safety protocols can be seen in various maritime operations. For example, using drones for overboard search and rescue operations, or employing robust communication systems, ensuring that vessels remain in constant touch with on-shore bases and other vessels, fostering a network of support and assistance in the vast ocean expanses.

Legal and Ethical Implications

Implementing technology in maritime operations inevitably encounters a sea of legal aspects. From adhering to international maritime laws, which might dictate specifications around vessel equipment and operation, to ensuring compliance with data protection regulations when utilizing data-centric technologies – navigating through these legal waters is complex and critical. This also entails keeping abreast of ongoing and emerging regulations, ensuring that all technological deployments are future-proofed against upcoming legal frameworks.

Ethical Considerations of Technological Implementations

On the ethical front, deploying technology in maritime operations must consider the potential social and human impacts. Ensuring that technology does not marginalize or endanger the human workforce is pivotal. This involves creating robust transition and training plans, ensuring that the existing workforce is adeptly upskilled to navigate and operate within the newly technologized environment. Moreover, ensuring that technological applications, particularly AI, operate within a framework that is transparent, unbiased, and fair is critical to maintaining ethical integrity.

A holistic, informed, and cautious approach is needed to navigate risk management, safety, legality, and ethics in this technological journey in the maritime domain and ensure that innovation propels the industry forward without sinking the ship of responsibility and human-centricity.

Analysis and Anticipated Developments

For a successful and sustainable future, the maritime sector must integrate future technical advancements with strategic planning. Anticipating autonomous ships, advanced AI-powered analytics, blockchain for secure and transparent transactions, and increased IoT for integrated maritime ecosystems presents problems and opportunities. Consequently, robust policies must carefully combine technological expertise with agility to make the sector resilient and responsive to technology advances. These technological advancements and strategic implementations must be grounded in ethics to avoid social, economic, and environmental impacts, ensuring that the technological voyage balances humanitarian and ecological concerns.

Modern maritime leaders must navigate the future with foresight, ethics, and a commitment to integrating technological advancements with societal and industry wellbeing to ensure it is advanced, equitable, sustainable, and deeply human-centric.

Chapter 6

THE GREEN ENERGY IMPERATIVE

International Commitments and Climate Agreements

The shipping business, similar to several other sectors, is presently functioning within the framework of diverse international accords aimed at mitigating the environmental consequences of human actions. The Paris Agreement states that temperature goals require emissions to fall by one-third or one-half by 2030 and to be zero by 2040 or 2050. It has been adopted by 189 out of 197 nations seeking to mitigate global warming. These companies have restricted the increase in average world temperature to a level much below 2 degrees Celsius compared to pre-industrial levels.

The maritime sector is dedicated to reducing greenhouse gas emissions. Compared to 2008, The International Maritime Organization (IMO) plans to cut greenhouse gas pollution by at least half each year by 2050. These efforts show the decreasing side of our environmental situation and how the world works together to fix it. Currently, the maritime industry is adjusting its strategies to align with international agreements, focusing on implementing cleaner operations.

Environmental Impact of Traditional Maritime Operations

History shows that the international emissions of carbon dioxide (CO2), nitrogen oxides (NOx), and sulfur oxides (SOx) are largely caused by the maritime business. Previously, ships used heavy fuel oil that used to burn and release harmful pollutants into the Earth's atmosphere. Furthermore, apart from air emissions, conventional shipping operations are also associated with oil spill incidents, ballast water discharge, and several other types of marine pollution. These practices resulted in severe repercussions for marine ecosystems, such as the deterioration of coral reefs, negative impacts on aquatic organisms, and the exacerbation of climate change on a larger scale. The environmental consequences associated with these conventional practices prompted an expedited demand for a more ecologically friendly maritime sector.

Stakeholder and Consumer-Driven Demand for Sustainable Practices

The demand for sustainability has evolved beyond being solely a regulatory or environmental concern and has now assumed a significant role in the commercial realm. Stakeholders, encompassing investors, shareholders, and partners, are progressively appraising organizations with regard to their Environmental, Social, and Governance (ESG) performance. It is acknowledged that there exists a positive relationship between sustainable operations and both long-term profitability and risk mitigation. There is an increasing consumer desire for ethical and sustainable behaviors. With the increasing accessibility of information, consumers are becoming more knowledgeable and discerning and frequently choose to patronize firms that are in line with their individual ideals. In the context of the shipping sector, there exists a significant imperative to showcase sustainable practices. This imperative is driven not only by the necessity to adhere to regulatory requirements but also by the need to adapt to changing market dynamics and consumer preferences.

The Immediate and Long-Term Benefits of Green Shipping

Embracing sustainability in the maritime sector is not merely an ethical obligation but also a strategic imperative. The shift towards green shipping brings forth numerous benefits, spanning environmental, economic, and social domains, showcasing that a sustainable approach is beneficial not just for the planet but also for the industry and society at large.

Environmental Benefits: Curbing Emissions and Marine Contamination

As the maritime industry gravitates towards green shipping, the environment stands to benefit immensely. Adopting sustainable practices leads to a significant reduction in the release of noxious emissions, notably carbon dioxide (CO_2), sulfur oxides (SO_x), and nitrogen oxides (NO_x). By switching to cleaner fuel sources and implementing energy-efficient technologies, ships can greatly diminish their carbon footprint. Moreover, with environmentally conscious shipping, the risk of marine pollution decreases. This includes reducing oil spills and minimizing the discharge of ballast water and other pollutants, thus helping protect fragile marine ecosystems and biodiversity.

Economic Advantages: Cost-Efficiency and Future-Proofing Operations

Green shipping is environmentally responsible, as well as it offers tangible economic benefits. In the short term, energy-efficient systems can lead to fuel savings, translating to reduced operational costs. Furthermore, as global regulations tighten around environmental standards, companies that have already adopted sustainable practices will face fewer disruptions and potential penalties. They are essentially future-proofing their operations against stringent regulations that might be introduced down the line. Additionally, as the market becomes more environmentally conscious, companies prioritizing green shipping can differentiate themselves, potentially garnering more business and establishing a reputable brand image.

Social Implications: Ensuring Health and Boosting Public Image

The transition to sustainable maritime practices has profound social implications as well. Reduced emissions from ships contribute to cleaner air, directly benefiting communities living near ports and coastlines. This can lead to fewer health issues associated with air pollution, such as respiratory ailments. Moreover, as public awareness about environmental concerns grows, companies championing green shipping practices can expect a favorable public perception. Being viewed as a responsible and forward-thinking enterprise can lead to positive public relations, increased customer loyalty, and, potentially, a larger market share.

Unveiling Alternative Fuels and Propulsion Systems

As the maritime industry grapples with the pressing need to reduce its carbon footprint, the exploration and adoption of alternative fuels and innovative propulsion systems emerge as crucial components. This shift not only resonates with global sustainability goals but also offers promising solutions for a cleaner, greener maritime future.

Evaluating Different Fuel Options

The quest for sustainability has prompted the maritime industry to assess and experiment with diverse fuel options. Each alternative carries its own set of advantages, challenges, and potential, necessitating thorough exploration to discern its feasibility and long-term viability.

Biofuels and Their Feasibility

Biofuels, derived from organic matter, present a renewable energy source for the maritime sector. They can potentially reduce greenhouse gas emissions significantly when compared to traditional fossil fuels. Additionally, biofuels can be integrated into existing ship engines with minor modifications, making them a relatively accessible alternative. However, concerns regarding their production scale, cost, and competition with food resources highlight the need for sustainable sourcing.

LNG (Liquefied Natural Gas) as an Alternative

LNG is increasingly being seen as a promising transition fuel for the maritime sector. With lower sulfur, CO_2, and NO_x emissions compared to heavy fuel oil, LNG addresses several environmental concerns associated with conventional maritime fuels. Infrastructure for LNG refueling is also gradually expanding, yet challenges remain regarding methane slip and long-term sustainability.

Hydrogen Fuel Cells and Potential Applications

Hydrogen, when used in fuel cells, produces electricity through a chemical reaction, emitting only water vapor as a byproduct. This makes hydrogen fuel cells a compelling option for zero-emission maritime operations. Current research focuses on improving the efficiency, storage, and safety of hydrogen systems. Though in nascent stages for large-scale maritime applications, the potential of hydrogen as a clean fuel source is undeniable.

The analysis of the possibilities of these alternative fuels reveals that the maritime industry is on the verge of a significant period of transformation. The adoption of these innovations not only facilitates the implementation of sustainable practices but also establishes the industry as a leader in global environmental conservation.

Innovations in Propulsion Techniques

In the modern maritime era, propulsion techniques have transcended the conventional, venturing into realms that amalgamate innovation with sustainability. The industry's pivot to greener solutions has birthed an array of propulsion methods that promise efficiency while treading lightly on the environment.

Hybrid Systems Combining Traditional and Electric Propulsion

Hybrid propulsion systems, as the name suggests, combine the strengths of traditional propulsion mechanisms with electric propulsion. These systems allow ships to operate efficiently, often utilizing electric power during periods of low power demand and switching to conventional propulsion during higher demands. The result is a significant reduction in emissions, especially in coastal regions or ports where emission regulations are stringent.

Solar and Wind-Assisted Propulsion Solutions

Harnessing the natural energies of the sun and wind, some modern ships are integrating solar panels and wind turbines into their energy mix. Solar panels affixed on ships can generate electricity to power various onboard systems. Similarly, rotor sails or Flettner rotors can be installed on ships to harness wind energy, aiding propulsion and reducing the reliance on fossil fuels. These solutions, while supplemental, contribute to reducing the carbon footprint of maritime operations.

Magneto hydrodynamic (MHD) Propulsion and Other Emergent Technologies

MHD propulsion represents a cutting-edge technology where seawater, when passed through a magnetic field, acts as the electrode and propels the ship forward. It's a clean, noiseless, and efficient system with no moving parts, eliminating mechanical wear and tear. Though still in developmental stages and faced with challenges like high energy consumption, MHD propulsion offers a glimpse into the potential future of maritime transportation.

As the marine industry innovates, propulsion methods show its dedication to efficiency and sustainability. The horizon suggests a day when ships glide effortlessly, driven by a blend of conventional and cutting-edge means for environmental protection.

Case Study: MSC's Pursuit for Sustainable Shipping

As industries worldwide spindle towards sustainable operations, the maritime sector witnesses exemplars in the form of leading companies setting the green course. Mediterranean Shipping Company (MSC), one of the world's largest shipping lines, has emerged as a trailblazer, earnestly pursuing a path of eco-friendly maritime operations.

MSC's Sustainability Initiatives

MSC has long recognized the pressing need for sustainability in the maritime industry. Grounded in a commitment to the planet and future generations, the company has continually revamped its operational ethos to resonate with environmental stewardship.

MSC's Environmental Strategy

MSC's environmental strategy is robust, holistic, and attuned to global sustainability goals. Recognizing the crucial role shipping plays in global trade and its consequent environmental impact, MSC is driven by a responsibility to minimize this footprint. Their strategy encompasses reducing emissions, sustainable ship recycling, enhancing energy efficiency, and ensuring responsible operations across their vast fleet.

Specific Goals and Targets Set by MSC

In alignment with international goals and the broader vision of decarbonizing the shipping industry, MSC has set ambitious targets for itself. Key among them is the commitment to a 30% reduction in CO_2 emissions per transport work by 2030, a significant stride towards the long-term aspiration of complete decarbonization. Additionally, MSC is actively investing in research and development, aiming to explore and adopt alternative fuels and advanced green technologies, underscoring their dedication to ushering in a new era of sustainable shipping.

MSC's dedicated pursuit of sustainable shipping offers a blueprint for the maritime industry, showcasing the viability of balancing efficient operations with environmental responsibility. Their initiatives not only cement their position as industry leaders but also inspire a wave of green transformation across the maritime spectrum.

MSC's Adoption of Alternative Fuels and Propulsion

Embarking on a journey towards sustainable shipping, MSC's dedication to adopting green technologies and fuels is evident. Their commitment underscores the fact that sustainable operations can be amalgamated with commercial viability, setting benchmarks for others in the sector.

Use of Biofuels and Other Green Fuel Solutions

MSC has actively explored and adopted the use of biofuels as a viable green solution for their fleet. Partnering with trusted suppliers, they have commenced the blending of sustainable, second-generation biofuels in their regular marine fuels. This not only reduces the carbon footprint but also decreases other emissions like sulfur oxides. Beyond biofuels, MSC is researching and adopting other green fuel solutions, displaying a multi-pronged approach to environmental conservation.

Investment in Green Technology for Vessels

Recognizing that fuel is just one component of sustainable operations, MSC has heavily invested in state-of-the-art green technologies for their vessels. Their ships are being equipped with energy-saving appliances, improved hull designs for better hydrodynamics, and advanced waste treatment systems. Moreover, they are incorporating air lubrication technology for reduced friction and exhaust gas cleaning systems to cut down emissions. All these innovative adoptions are a testament to MSC's holistic approach to green shipping.

MSC's unwavering commitment to sustainable maritime operations, primarily through the adoption of alternative fuels and green technologies, shines as a beacon for the entire industry. Their initiatives resonate with the imperative that environmental conservation and commercial operations are not mutually exclusive but can coexist harmoniously, steering the future of shipping.

Lessons from MSC: Best Practices and Challenges

Delving into the story of MSC's green initiatives provides the maritime sector with critical lessons in sustainable transformation. Their endeavors showcase how even large-scale operations can pivot towards more environmentally friendly practices, offering invaluable insights for others.

Key Takeaways from MSC's Green Journey

Following are the key takeaways from MSC's Green Journey:

Strategic Vision

One of the foremost lessons from MSC is the importance of a clear and long-term vision for sustainability. It's not just about short-term adaptations but understanding the broader environmental outlook and setting milestones accordingly.

Stakeholder Engagement

MSC's green journey was not an isolated endeavor. They actively engaged with stakeholders, from suppliers to customers, ensuring that sustainability became a shared objective.

Continuous Innovation

MSC's commitment to research and development stands out. Rather than relying on existing solutions, they invested in creating and adapting newer technologies, ensuring their fleet remained at the forefront of green shipping.

Monitoring and Feedback

Regular monitoring and feedback mechanisms ensured that MSC could track their progress, recalibrate their strategies, and maintain transparency with their stakeholders.

Challenges Faced and Strategies for Overcoming Them

MSC adopted the following strategy to overcome their challenges:

High Initial Costs

The shift towards sustainable shipping comes with its financial burdens. MSC addressed this by viewing it as a long-term investment, ensuring they would reap the benefits in future savings and brand reputation.

Technological Hurdles

Implementing new technologies often brings unforeseen challenges. MSC dealt with these by partnering with experts and ensuring their staff received adequate training.

Supply Chain Complications

With the introduction of new fuels and technologies, ensuring consistent supply chains was a challenge. MSC navigated this by strengthening ties with suppliers and diversifying their supply sources.

Regulatory Adjustments

As the maritime sector evolves, so do the regulations governing it. MSC maintained a proactive stance, staying ahead of regulatory changes and often exceeding the minimum requirements.

MSC's green journey, rife with both achievements and challenges, stands as a testament to the fact that with commitment and strategic planning, sustainable maritime operations are achievable. Their story offers both inspiration and practical lessons, underlining the notion that sustainability is not a mere trend but an imperative for the future of the maritime industry.

The Future of Green Shipping

The maritime sector stands on the brink of a revolutionary transformation driven by the dual forces of environmental necessity and technological innovation. As the world rallies towards a sustainable future, the seas are bound to witness an unprecedented wave of green transitions.

Emerging Technologies in Alternative Fuels

Algae-Based Biofuels

Marine microalgae are gaining traction as a potential feedstock for biofuels. They proliferate, require minimal land, and can be converted into various types of fuels.

Ammonia as Fuel

Preliminary research indicates that ammonia when burned in the presence of oxygen, releases energy without emitting CO_2. Its potential as a carbon-free fuel source is actively being explored.

Methanol from Renewable Sources

Methanol is another contender in the alternative fuels race. When sourced from renewable methods, it presents an environmentally friendly option with significant calorific value.

Potential Breakthroughs in Propulsion Systems

Fully Electric Propulsion

Building upon the hybrid systems of today, the future might witness ships powered entirely by electricity sourced from renewable grids or onboard energy generation methods.

Nuclear Marine Propulsion

Though fraught with controversy, small modular reactors designed explicitly for marine use could provide a long-term, low-emission propulsion solution.

Sail-Assisted Propulsion

A blend of the old and new - integrating advanced materials and designs with the age-old concept of sails, potentially reducing dependency on fuels, especially during favorable weather conditions.

The horizon of green shipping gleams with potential. As technology and innovation continually evolve, the maritime sector is poised to redefine its relationship with the environment. The ships of the future, powered by alternative fuels and groundbreaking propulsion techniques, will not only sail the seas but also safeguard them for upcoming generations.

Strategies for Companies to Transition towards Sustainability

In the emergence of a new maritime era, sustainability is more than an environmental responsibility; it's a strategic imperative. Companies at the helm need a clear roadmap to navigate the challenges and harness the opportunities of this green transition.

Financial and Operational Planning for Sustainable Transitions

Investing in Research and Development (R&D)

Dedicate resources to R&D to foster innovations tailored for the company's operations. Exploring in-house solutions or adapting existing technologies can provide a competitive advantage.

Budgeting for Sustainable Infrastructure

Retrofitting older vessels or infrastructure to meet green standards can be capital-intensive. Financial planning, including considerations for grants, subsidies, or green bonds, is crucial.

Operational Shifts

Companies might need to reconsider routes, speeds, or schedules to align with sustainable operations. These decisions, while possibly affecting short-term profitability, are investments in long-term viability and reputation.

Partnering with Green Tech Providers and Research Institutions

Collaborations for Custom Solutions

By collaborating with tech providers, companies can develop solutions tailored to their unique operational challenges. These partnerships ensure that the technology aligns seamlessly with the company's maritime operations.

Engaging with Academic Institutions

Universities and research institutions are often at the forefront of innovation. Partnerships can provide access to the latest research, potential pilot projects, and even future experts in the field.

Participating in Industry Consortia

Joining forces with peers can help in sharing knowledge, pooling resources, and even lobbying for favorable policies or regulations. Collective efforts can accelerate the industry's sustainable transformation.

Sustainable maritime practice is a journey, not a destination. As organizations strategize their direction, the utilization of strategic foresight, in conjunction with collaborative engagements, will serve as their most invaluable guiding tool. In this joint project, each marine mile traveled in an environmentally responsible manner will have a cascading effect, leading to a more balanced and interconnected future for the Earth.

Collaborative Efforts: How Alliances Can Propel Green Shipping

In the expansive maritime horizon, collaboration acts as the beacon, illuminating pathways to sustainable transformations. United by shared objectives, alliances can amplify efforts, making the dream of a green maritime composition an attainable reality.

Role of International Collaborations and Alliances

Pooling Resources

International collaborations offer a platform for members to pool resources, both intellectual and financial, to tackle the challenges of green shipping. This can lead to breakthroughs in technology, streamlined operations, and financial synergies that might be out of reach for individual entities.

Standardizing Practices

Global alliances can help in formulating universal standards and best practices. With a unified approach, the maritime industry can ensure consistent and effective green implementations across regions.

Policy Advocacy

A collective voice is powerful. International collaborations can play a pivotal role in lobbying for favorable environmental regulations, subsidies, and incentives that can accelerate the shift toward sustainable shipping.

Case Examples of Successful Green Alliances in the Maritime Sector

Getting to Zero Coalition

A notable initiative that aspires to make zero-emission vessels a commercially viable option by 2030. Comprising members from maritime, energy, infrastructure, and finance sectors, it exemplifies how diverse stakeholders can collaborate for a shared goal.

Trident Alliance

Focused on robust enforcement of sulfur regulations, this coalition of shipping owners and operators has been instrumental in ensuring compliance and advocating for fair competition in the sector of sustainable maritime practices.

The maritime sector emphasizes that sustainability is a communal effort by leveraging solidarity. While vessels navigate towards more environmentally sustainable shores, driven by the ideals of cooperation, they forge a course brimming with potential for both the maritime sector and the global community.

Chapter 7

ECONOMIC FORECAST
AND BUSINESS MODELS

Maneuvering through the ebb and flow of the global economic tides, the maritime industry finds itself at the forefront of countless transformative changes. As the backbone of global trade, understanding the economic path of this sector is essential for stakeholders within and for ancillary industries, policymakers, and international trade partners.

Economic Projections for the Next 5-10 Years

Anticipated Growth Regions and Potential Economic Hotspots

The maritime sector, a cornerstone of global commerce, is rapidly evolving. With shifting economic tides, certain regions are emerging as the epicenters of growth, and this, in turn, is reshaping global trade dynamics.

Emerging Markets in Southeast Asia

Countries such as Indonesia and Vietnam, already key players in the global maritime sector, are predicted to play even more prominent roles in the near future. Both nations are investing heavily in port infrastructure, seeking to accommodate larger vessels and increase cargo handling capabilities. Indonesia, with its strategic location and archipelagic advantage, aims to transform itself into a global maritime fulcrum. Vietnam, benefiting from the ongoing US-China trade tensions, has seen a diversion of manufacturing bases to its shores, thereby increasing its maritime trade volumes.

Africa's Maritime Potential

Africa, with its vast coastline and strategic geographical position, is another region brimming with maritime potential. Countries like Nigeria, South Africa, and Kenya are heavily investing in their ports, aiming to become the transit and logistics hubs of the continent. Nigeria's Lekki deep sea port, for example, is anticipated to be one of Africa's deepest ports, capable of handling even the largest vessels. Moreover, with China's Belt and Road Initiative extending its tendrils into the African continent, there is a pronounced increase in maritime activities in the region.

The Northern Sea Route's Potential

As the global climate changes and Arctic ice recedes, the Northern Sea Route offers a tantalizing opportunity. This potential new commercial maritime corridor could drastically reduce transit times between Europe and East Asia. While there are environmental and geopolitical concerns associated with this route, its economic potential is undeniable. Russia, in particular, is keen on developing this route, viewing it as a strategic advantage that could shift a significant portion of global trade traffic.

The maritime sector, characterized by its dynamic nature, is currently undergoing a significant period of transformation. As emerging markets in Southeast Asia and Africa rise in prominence and previously inaccessible routes like the Northern Sea Route come into play, the world map of maritime commerce is being redrawn. For stakeholders in the industry, understanding these shifts and positioning themselves strategically will be vital to harness the upcoming tidal wave of opportunities.

Impact of Geopolitical Scenarios and Trade Policies

The very nature of maritime trade means it intersects with international politics at almost every turn. Ships ferrying goods from one part of the globe to another act as bridges between economies and cultures, but they also find themselves at the mercy of shifting geopolitical winds.

Regional Self-Reliance and Supply Chain Diversification:

The COVID-19 pandemic exposed vulnerabilities in the global supply chain, prompting nations to rethink their trade strategies. Many are emphasizing regional self-reliance, aiming to reduce dependency on singular, major trade partners. This move might lead to a more distributed pattern of trade, with countries exploring newer markets and forging partnerships with nations they hadn't prioritized before. Such a shift not only provides a cushion against future disruptions but also distributes economic growth more evenly.

Trade Embargoes and Sanctions

These are potent tools in the arsenal of geopolitical maneuvering. As seen in the past, sanctions or embargoes can severely affect the maritime trade of a country, redirecting trade flows and causing significant financial ramifications. Ports can find themselves suddenly less busy, and shipping lanes might change overnight, underscoring the industry's vulnerability to political decisions.

US-China Trade Tensions

The trade tensions between these two global superpowers, involving tariffs and counter-tariffs, have ripple effects across the world. The maritime sector, as a primary medium of trade between these giants, feels these effects profoundly. Vessels have been rerouted, and trade volumes have shifted as manufacturers look for alternatives to avoid tariffs, leading to opportunities for countries like Vietnam and Mexico.

Brexit Implications

The decision by the United Kingdom to withdraw from the European Union resulted in a multitude of uncertainties for the marine industry. The negotiation of new trade agreements and reassessing current ones presents a range of difficulties and opportunities. With these forthcoming changes in trade patterns following Brexit, ports such as Rotterdam, Antwerp, and Le Havre are undertaking strategic adjustments to optimize their operations.

RCEP and Regional Trade Agreements

The RCEP, a free trade agreement spearheaded by ASEAN countries and their partners, is the largest in the world by GDP. It aims to streamline existing agreements and create a unified trading bloc. For the maritime industry, this could mean increased trade flows within Asia and with partners like Australia and New Zealand, altering vessel traffic patterns and port dynamics.

The nexus of geopolitics and maritime trade is complicated and ever-evolving. While geopolitical events can introduce uncertainty and risks to the industry, they also present opportunities for those agile enough to adapt. By staying informed and prepared, maritime stakeholders can navigate these turbulent waters and even harness them for growth.

Expected Shifts in Major Trade Routes and Maritime Corridors

Trade routes, historically, have always been fluid, adapting to the times based on socio-political, economic, and technological factors. In the current era, a confluence of geopolitics, infrastructure projects, and the push for sustainability is reshaping these crucial arteries of global commerce.

Belt and Road Initiative (BRI)

China's ambitious Belt and Road Initiative is a sprawling network of trade routes encompassing land and sea, seeking to connect China with Europe, Africa, and other parts of Asia. Maritime routes, particularly the 21st Century Maritime Silk Road, aim to connect China's coastal regions to Europe through the South China Sea and the Indian Ocean. This endeavor might see ports from Southeast Asia to East Africa witnessing increased traffic and investments. On land, the initiative comprises a network of railways, roads, pipelines, and utility grids stretching from East Asia to Europe. This new Silk Road is poised to enhance connectivity, reduce transit times, and shift the epicenters of trade.

Eco-Efficient Routes

The global push towards sustainability is now affecting route selection for shipping. Shorter, more direct routes not only save time but also reduce carbon emissions. Additionally, with the advancement of technology, ships can use real-time data to find the most efficient path, avoiding stormy weather or leveraging ocean currents to save fuel. These innovations could redefine preferred maritime corridors.

Northern Sea Route (NSR)

The melting of Arctic ice, though an alarming indicator of global warming, is opening up the Northern Sea Route. This route connects Northeast Asia to Europe, offering a much shorter path than the traditional Suez Canal route. The NSR can reduce voyage times between Asia and Europe by up to 10 days, translating to lower fuel consumption and reduced emissions. As it becomes more navigable, it's likely that traffic along this route will increase, especially for eco-friendly vessels designed to navigate these icy waters.

Infrastructure Development and Trade Routes

Large-scale infrastructure projects, whether they're deep-sea ports, canals, or rail networks, can dramatically alter trade dynamics. A new port can shift cargo traffic away from congested ports, while a canal can create shortcuts, making long detours unnecessary. These developments can redefine maritime corridors and reshape the economic prospects of entire regions.

The current state of trade routes, which serve as crucial conduits for global business, is characterized by a dynamic and ever-changing nature. The entities in question are shaped by the influence of geopolitics, crafted through the development of infrastructure, and enhanced through the crucial perspective of sustainability. It is advisable for progressive enterprises and players in the marine industry to closely observe these changes, modify their approaches, and establish advantageous positions within this dynamic environment.

As the following decade approaches, the marine industry is on the cusp of significant economic transformations. These changes, influenced by a combination of regional growth prospects, global geopolitics, and transformative projects, necessitate an agile, informed, and strategic approach from maritime stakeholders. Ensuring resilience and adaptability amongst these shifts will secure profitability and solidify a position in the vanguard of global maritime trade.

Influence of Global Market Trends

As global market trends transform, so too does the maritime industry, adapting its sails to the winds of change. The relationship between burgeoning e-commerce, increasingly stringent shipping regulations, and unpredictable market fluctuations creates a maelstrom of challenges and opportunities for maritime operations. To steer successfully through these turbulent waters, a deep understanding of these influences becomes imperative.

The Growth of E-Commerce and its Ramifications for Shipping

The digital marketplace revolution, symbolized by e-commerce, has not just changed how consumers shop but has also significantly impacted the maritime industry's operational dynamics.

Consumer Expectations and Demand

The culture of "buy now, get it tomorrow" has driven shipping companies to innovate for quicker turnaround times. This immediacy, while beneficial for consumers, places increased pressure on shippers to minimize transit times. Enhanced tracking systems and transparent cargo movement are now a necessity to meet these consumer expectations.

Customized Cargo Solutions

Unlike traditional bulk shipments, e-commerce often involves shipping a diverse array of products in smaller quantities. This demands more versatile cargo solutions, including specialized containers and packaging methods to ensure product safety and efficient space utilization.

Dynamic Port Operations

Ports, once accustomed to dealing with larger cargo ships coming in at predictable intervals, now have to manage a mix. They must handle smaller vessels that arrive more frequently alongside traditional cargo ships. This requires adaptive scheduling, increased dock efficiency, and investments in infrastructure to facilitate quicker loading and unloading processes.

Enhanced Warehousing and Distribution Centers

To fulfill the e-commerce promise of fast deliveries, strategically located warehouses and distribution centers have become vital. Close proximity to ports ensures that goods can be quickly moved to these centers, sorted, and dispatched for last-mile delivery.

Technological Integration

E-commerce thrives on technology, and the shipping industry has had to adapt in kind. From advanced tracking systems and automated warehousing to digital freight forwarding platforms, the convergence of technology and shipping ensures a seamless flow of goods from sellers to buyers.

Green Shipping for E-Commerce

As consumers become more eco-conscious, there's a demand for sustainable shipping solutions, even in e-commerce. Shipping lines are looking to reduce their carbon footprint, incorporate recyclable packaging, and ensure environmentally friendly last-mile delivery methods, like electric vehicles.

E-commerce has set forth a wave of changes, making adaptability the hallmark of modern shipping operations. The shipping industry's response to e-commerce's demands, from ports to last-mile delivery, underscores its commitment to evolving and serving the new-age consumer.

Economic Repercussions of Stringent Shipping Regulations

In the bid to foster a more sustainable and environmentally-conscious maritime industry, international bodies have been introducing a range of stringent shipping regulations. Though these rules champion a noble cause, their implementation carries significant economic implications for various stakeholders.

Cost Implications

Introducing new technologies or switching to cleaner fuels often means considerable upfront capital investments for shipowners. Retrofitting vessels with new equipment or emission-reducing technologies, such as scrubbers, demands both time and money. In the short term, these expenditures can weigh heavily on operators, especially those with large fleets.

Market Dynamics and Fuel Prices

The push towards low-sulfur fuels, driven by regulations like the IMO 2020, has caused fluctuations in the demand and pricing of different types of bunker fuels. As the demand for cleaner fuels rises, prices could soar, further straining the already tight margins of shipping operators.

Operational Challenges

Adhering to new regulations often means retraining crew, revising operational protocols, and ensuring strict compliance during voyages. Non-compliance could lead to hefty fines, detentions, or, in some cases, vessel seizures, all of which carry heavy economic consequences.

Opportunities for Green Innovations

On the other side, the push for sustainability has spurred research and development in green maritime technologies. Companies pioneering in these areas stand to gain significant market share. Clean energy startups, sustainable shipbuilding firms, and eco-friendly supply chain solutions providers are emerging as key players in this new maritime sector.

Long-term Economic Benefits

While initial adaptation might be economically challenging, in the long run, sustainable practices can offer cost savings. Reduced fuel consumption, increased operational efficiencies, and positive brand recognition can all contribute to better financial performance.

Insurance and Financing

Financial institutions and insurance companies are becoming increasingly cautious about their exposure to non-compliant or environmentally risky ventures. Ships adhering to the latest regulations might find it easier to secure financing or insurance at competitive rates, providing an indirect economic advantage. The marine sector is currently at a critical juncture as it travels the economic complexities associated with complying with new rules while striving to achieve sustainable practices' long-term advantages. Progressive organizations perceive these laws as obstacles and agents that propel the sector toward a future characterized by enhanced accountability and economic robustness.

Market Fluctuations and Their Effect on Maritime Operations

Volatile market conditions, driven by factors ranging from geopolitical tensions to global pandemics, can dramatically impact maritime operations. Fluctuating oil prices influence shipping costs, while shifts in consumer demand patterns can lead to vessel overcapacity or shortages. The recent Suez Canal blockage highlighted the domino effect a single disruption can have on global supply chains. To mitigate such challenges, maritime operations are investing in risk assessment tools, diversifying routes, and enhancing supply chain visibility.

The connection between the maritime industry and global market trends is complex and dynamic. Whether it's the transformative influence of e-commerce, the economic weight of regulatory compliance, or the unpredictability of market volatilities, maritime entities must remain agile and informed. Embracing these changes, rather than resisting them, will position the industry to invite emerging opportunities and ensure its continued relevance and resilience in the global economy.

Investing in the Future: Training and Technology

As the maritime industry evolves, entwined with rapid technological advancements and a shifting global prospect, the value of investing in both training and technology has never been more apparent. Equipping the workforce with the right skills and tools is not merely an investment in their personal growth but a strategic move to ensure the long-term viability and competitiveness of maritime enterprises.

The Imperative of Employee Training Programs

Role of Skill Enhancement in the Modern Maritime Sector

In an industry that's continually influenced by technological innovations, regulatory adjustments, and sustainability pressures, the emphasis on relevant skills is paramount. Modern maritime operations demand personnel adept not just in traditional seafaring skills but also in emerging areas like digital operations, environmental stewardship, and advanced vessel technologies. Skill enhancement ensures that the maritime workforce remains competent, versatile, and ready to tackle the challenges of the evolving seascape.

Benefits of Continuous Learning and Specialized Training

1. **Improved Operational Efficiency**: Well-trained employees, familiar with the latest tools and practices, tend to make fewer errors, leading to smoother operations.
2. **Enhanced Safety**: Continuous learning can significantly mitigate risks, reducing accidents and ensuring safer maritime operations.
3. **Employee Retention**: Offering training opportunities can boost employee morale, job satisfaction, and loyalty. When workers feel valued and see growth opportunities, they are less likely to seek employment elsewhere.
4. **Financial Upsides**: Over time, the financial returns of investing in training – through boosted productivity, reduced errors, and better decision-making – often outweigh the initial training costs.

Preparing the Workforce for Digital Transformation and Sustainable Operations

The marine industry is currently at the precipice of a transformative digital transformation. From AI-driven logistics to blockchain in supply chain transparency, digital tools are reshaping the industry. Furthermore, as the industry leans more towards sustainability, understanding eco-friendly operations is vital. Training programs need to focus on equipping the workforce with skills in digital tools, data analytics, green technologies, and sustainable best practices. Such initiatives ensure that employees are not only keeping pace with industry transformations but are also active contributors to driving change. Investing in employee training is a strategic decision that establishes a foundation for future success, transcending being a mundane expenditure. In a contemporary era characterized by the convergence of tradition and innovation within the marine industry, the search for ongoing education emerges as a guiding principle that directs firms toward advancement, financial success, and readiness to confront forthcoming problems.

Embracing Cutting-Edge Technologies

The maritime sector, historically seen as traditional and resistant to rapid change, is now at a critical juncture. As the global economy becomes more interconnected and digitized, the maritime industry is embracing a suite of cutting-edge technologies. These technologies are not just fads but transformative tools reshaping the very fundamentals of maritime operations, logistics, and management.

Adoption and Integration of AI, IoT, and Blockchain in Maritime

Artificial Intelligence (AI)

AI's influence is evident across various maritime facets. Predictive maintenance, powered by AI algorithms, can foresee potential equipment failures, thereby saving both time and resources. Furthermore, AI-driven route optimization tools can analyze weather patterns, currents, and other data points to suggest fuel-efficient and time-saving routes.

Internet of Things (IoT)

IoT devices have proliferated across ships, ports, and containers. These devices continuously collect and transmit data, allowing for real-time monitoring of vessel performance, cargo conditions, and even crew well-being. When integrated with AI, the data from IoT devices can lead to actionable insights, further streamlining operations.

Blockchain

Primarily known for its application in cryptocurrencies, blockchain offers a transparent, immutable, and secure method of recording transactions. In the maritime sector, this can translate to more transparent supply chains, ensuring all stakeholders have access to unalterable shipping data, enhancing trust, and reducing discrepancies.

Advancements in Digital Solutions for Logistics and Fleet Management:

Digital solutions have revolutionized how shipping companies manage their fleets and logistics. Modern fleet management systems allow operators to monitor vessel locations, performance, and health in real time. Digital twins, virtual replicas of physical vessels, can be used to run simulations, predicting how ships might react under various scenarios. On the logistics side, digital platforms are enabling better cargo tracking and inventory management and even automating some of the documentation processes, significantly reducing the administrative burden.

The Value of Data-Driven Decision-Making in Maritime Operations:

In today's maritime world, data is invaluable. With vast amounts of data being generated every second, from cargo temperatures to engine performance metrics, the industry is richer in information than ever before. When processed and analyzed correctly, this data can offer powerful insights. Data-driven decisions tend to be more accurate, timely, and aligned with both operational and strategic objectives. For instance, analyzing fuel consumption data across different routes can lead to more efficient routing decisions, ultimately resulting in cost savings.

Integrating cutting-edge technology into maritime operations is a competitive advantage, as it is rapidly becoming an industry standard. As technology continues to break boundaries, its role in defining the future of shipping is undeniable. Embracing these advancements is essential for maritime entities to ensure they remain relevant, efficient, and ready to direct the complexities of the modern global trade sector.

Adapting Business Models to Global Demands

In today's rapidly evolving maritime world, businesses must remain agile, frequently reassessing and adapting their models to align with global demands. With a growing emphasis on sustainability, ethical business practices, and efficient resource utilization, the maritime sector is seeing a paradigm shift in how business is conducted. One such influential concept that has gained significant traction is the Circular Economy Model.

The Circular Economy Model in Maritime

Principles and Benefits of a Circular Economic Approach:

At its core, the circular economy promotes resource maximization and waste minimization by ensuring products and materials remain in use for as long as possible. This model stands in stark contrast to the traditional linear economy, which follows a 'take-make-dispose' pattern.

In the maritime context, adopting a circular approach can have profound benefits:

1. **Resource Efficiency**: Circular practices such as refurbishing and recycling ship components can extend the lifecycle of materials, thereby reducing the need for extracting and processing new raw materials.
2. **Waste Reduction**: Instead of discarding old vessels, components can be reused, remanufactured, or recycled, significantly diminishing waste.
3. **Economic Benefits**: With efficient resource use comes cost savings. Moreover, adopting circular practices can open up new revenue streams, such as the sale of refurbished components or materials.

4. **Enhanced Sustainability**: Given the maritime industry's considerable environmental footprint, a circular approach can significantly mitigate negative impacts, aligning businesses with global sustainability goals.

Implementation Challenges and Examples of Successful Integration

Transitioning to a circular model in the maritime sector is not without challenges. These can range from initial investment costs to a lack of knowledge or technical expertise. Additionally, ensuring the entire supply chain aligns with circular principles can be daunting.

However, several players in the maritime space have successfully integrated circular principles. For instance:

1. **Ship Recycling**: Companies are increasingly focusing on green ship recycling, ensuring materials from decommissioned ships are reused or recycled, minimizing environmental harm.
2. **Remanufacturing Components**: Several maritime businesses are investing in remanufacturing operations, where components like ship engines are refurbished and resold, prolonging their lifecycle.
3. **Leasing Models**: Some maritime companies have started offering leasing models for equipment. Once the lease ends, the equipment is returned, refurbished, and leased again, promoting a continuous cycle of use.

The circular economy model, while challenging to implement fully, offers maritime businesses a path toward sustainable growth and operation. By realigning business practices with circular principles, maritime entities not only stand to reap economic benefits but also play a pivotal role in advancing global sustainability agendas. As global demands lean increasingly towards sustainable and efficient practices, the circular model is poised to be at the forefront of maritime business strategies.

Blockchain: Enhancing Transparency and Accountability

In an era where efficiency, security, and transparency are paramount, blockchain technology emerges as a guiding light, revolutionizing traditional maritime operations. Primarily known as the backbone of cryptocurrencies, blockchain's decentralized and immutable nature positions it as a powerful tool for enhancing transparency and accountability in the maritime sector's complex supply chains.

Fundamentals of Blockchain Technology in the Supply Chain

A blockchain is a digital ledger, recording transactions across many computers in such a way that any involved record cannot be altered retroactively without the alteration of all subsequent blocks. This ensures transparency and security.

When applied to supply chains in the maritime sector:

1. **Transparency**: Every transaction (or block) is visible to all parties involved, from the manufacturer to the end consumer. This visibility ensures authenticity and reduces the chances of fraud or tampering.
2. **Efficiency**: With blockchain, paper-based, time-consuming processes can be digitized, resulting in faster and smoother operations. Customs clearances, which traditionally take days, can be expedited by providing real-time access to necessary shipping data.
3. **Traceability**: Products can be traced back to their origin seamlessly. This is crucial for sectors like seafood or pharmaceuticals, where understanding the origin and journey of a product is vital for safety and compliance.

Case Studies: Companies Leveraging Blockchain for Streamlined Operations:

1. **Maersk and IBM's TradeLens**: In a bid to digitize the global shipping industry, Maersk and IBM launched TradeLens, a blockchain-enabled shipping solution. TradeLens provides real-time access to shipping data and shipping documents, including IoT and sensor data, ranging from temperature control to container weight.
2. **CargoX**: This company introduced the world's first blockchain Bill of Lading. Through blockchain, they ensure quick, safe, and cost-effective processing of Bills of Lading, dramatically reducing operation times.

3. **Walmart and IBM's Food Trust Solution**: While not exclusive to maritime, this collaboration uses blockchain to improve traceability in the food supply chain. Walmart can now trace the origin of over 25 products from five different suppliers in mere seconds rather than days.

Blockchain technology, with its potential to instill unparalleled transparency and efficiency, is steadily solidifying its place in maritime operations. By adopting and integrating blockchain, maritime businesses can achieve streamlined operations and bolster trust among all stakeholders. As the maritime industry becomes more digital, the role of blockchain is bound to be important, reshaping industry norms for the better.

Strategies for Navigating the Evolving Maritime Industry

The maritime sector is entering unknown territory with rapidly evolving technology and fluctuating global economic tides, so businesses need to be prepared with cutting-edge equipment and innovative approaches. Combining proactive adaptation with predictive analysis is essential for navigating this dynamic environment and allowing organizations to not only weather the storm but also capitalize on its winds of change.

Predictive Analysis and Proactive Adaptation
Using Data Analytics to Anticipate Market Changes:

Data analytics has become an invaluable compass for maritime businesses. By sifting through vast amounts of data—ranging from global economic indicators to specific shipping routes' performance—companies can discern patterns and trends. Advanced algorithms, coupled with historical data, allow businesses to forecast potential market shifts, customer preferences, and even probable logistical bottlenecks. For instance, analyzing data on cargo waiting times at specific ports can help shipping companies optimize schedules and reduce idle times.

Strategic Planning for Future Uncertainties and Opportunities

Anticipating changes is only half the battle; the real advantage lies in proactive adaptation. Once businesses have a forecast in hand, they must mold their strategies to align with these insights. This might involve diversifying shipping routes, investing in next-gen technologies, or retraining the workforce to handle emerging challenges. For instance, if data suggests a potential surge in demand in a specific region, companies can allocate resources and strengthen their presence there ahead of time. Conversely, predictive insights might signal an impending downturn, allowing businesses to hedge against risks and secure their assets.

Despite the turbulent currents in the sector, maritime enterprises can confidently chart their route by utilizing predictive analysis and adopting proactive adaptation. For businesses looking to lead the marine industry, the capacity to foresee and seize new possibilities and challenges is becoming more than just a useful skill.

Building Collaborative Alliances

In the marine industry, it is imperative to acknowledge that no company operates in isolation due to the vast and diverse nature of the sector. Forming strategic alliances and partnerships becomes essential, allowing businesses to pool resources, share knowledge, and jointly navigate the turbulent waters of global commerce. Collaborative alliances amplify both reach and capabilities, laying down a path for mutual growth. But how do these partnerships manifest, and what makes them effective?

The Role of Partnerships in Enhancing Business Reach and Capabilities

Partnerships in the maritime sector often provide more than just immediate transactional benefits; they pave the way for long-term growth and stability. By collaborating with others, businesses can access new markets, harness complementary skills, and even share the risks associated with larger investments or ventures. For instance, a shipping company might form an alliance with a tech firm to leverage cutting-edge fleet management systems, resulting in optimized operations and reduced costs.

Case Studies of Effective Maritime Collaborations for Mutual Growth

1. **Maersk and IBM**: One notable partnership in the maritime world is between shipping giant Maersk and technology behemoth IBM. The two industry leaders came together to create TradeLens, a blockchain-driven platform designed to promote more transparent and efficient global trade. By sharing data in real-time across the entire supply chain, TradeLens reduces

paperwork, speeds up customs clearance, and ensures all stakeholders have access to the information they need.

2. **CMA CGM and MSC Joining TradeLens**: Building upon the earlier case, two other major shipping companies, CMA CGM, and MSC, decided to join the TradeLens platform, reinforcing the idea that collaboration—even among competitors—can yield significant benefits. Their participation expanded the platform's reach, covering nearly half of the world's container cargo.

Building collaborative alliances in the maritime industry stands as a testament to the adage, "Together, we achieve more." These partnerships, rooted in mutual respect and a shared vision, not only propel individual companies forward but also catalyze innovations and advancements that move the entire industry toward a brighter, more interconnected future.

In the dynamic marine industry, innovative business models serve as symbols of transformative potential. Insightful economic planning and flexibility pave the way to industry success, illustrating the complex web of connections. Maritime enterprises need to be flexible, look forward, and stay steadfastly dedicated to a path of constant evolution now more than in the past. As the tides of change ebb and flow, those who seize the controls with conviction and foresight will maneuver through the waters of opportunity with unmatched prowess.

Chapter 8

CASE STUDIES: LEADING THE CHANGE IN GLOBAL MARITIME

In this era of technological advances and evolving global demands, the maritime industry stands at the crossroads of transformation. Despite the ebb and flow of global trade, certain corporations rise above the tide, guiding the industry with their visionary practices. These trailblazers, the pioneers of modern maritime, not only adapt to change but also spearhead it, setting benchmarks for the entire sector.

The Significance of Industry Leaders in Setting Maritime Trends

The maritime industry, with its vast expanse and intricacies, relies heavily on its behemoths to navigate uncharted waters. Industry leaders, such as Maersk, MSC, and CMA CGM, play a pivotal role in setting maritime trends. Their expansive operations, encompassing diverse geographies and cultures, provide them with unique insights into global demands and challenges. By championing innovative solutions, they pave the way for newer, more sustainable, and efficient maritime practices. Their influence transcends their immediate operations; their decisions often ripple across the industry, influencing policies, operational standards, and even international maritime regulations. In essence, by observing these leaders, one can predict the path of the entire maritime sector.

The Need for Studying Exemplary Businesses in an Evolving Sector

In a dynamic environment like the maritime industry, stagnation is not an option. For businesses, small or large, understanding the modus operandi of industry leaders is not just beneficial—it's crucial. Studying exemplary businesses provides a blueprint for success. This publication provides valuable perspectives on optimal methodologies, approaches to minimize risks, and groundbreaking solutions that have undergone rigorous testing and demonstrated significant efficacy on a large scope. Moreover, it offers an in-depth understanding of the future course of the sector, enabling companies to foresee approaching alterations and adjust their strategies in a proactive manner. By investigating deeply into the operations and strategies of leading maritime companies, businesses can gain a competitive edge, ensuring their longevity and success in an ever-evolving prospect.

The current marine sector leaders play a significant role beyond their commercial success, serving as a guiding force for many. Their study is not only intriguing but also crucial for individuals invested in the future of the maritime world, as their efforts, tactics, and advancements establish a foundation for succeeding developments.

Analytical Insights: Understanding the Big Players

The maritime industry is vast, diverse, and ever-evolving. Amidst its vast expanse, certain companies consistently emerge as industry forerunners, setting standards and charting new courses for others to follow. Understanding the intricacies of these major players—Maersk, MSC, and CMA CGM Group—requires a meticulous analytical approach. This not only uncovers the secrets behind their successes but also illuminates the broader trajectories of the maritime industry.

Criteria for Selecting Maritime Companies for Case Analysis

Choosing maritime giants for an in-depth case analysis requires a set of stringent criteria to ensure that their operations, strategies, and outcomes provide valuable insights into the industry's current state and future:

Market Presence

One of the primary criteria is the company's global footprint. Firms with extensive operations across multiple continents, such as our chosen trio, have a profound impact on international trade patterns and maritime trends.

Innovation and Adaptability

The maritime industry is in a state of flux, with technological advancements and sustainability demands constantly reshaping its sector. Companies that are at the forefront of innovation, as seen in Maersk's digital solutions or CMA CGM's sustainability initiatives, serve as excellent study subjects.

Financial Performance

A consistent track record of financial stability and growth, despite market fluctuations, is a clear indicator of a company's robust strategies and operational efficiency.

Stakeholder Engagement

Companies that actively engage with their stakeholders—be it customers, employees, or regulators—and have a reputation for transparency and corporate responsibility are prime candidates for analysis.

Industry Recognition

Awards, certifications, and recognitions by industry bodies and associations are testimony to a company's contributions and influence in the maritime sector.

The Metrics to Gauge Influence and Success in the Industry

To holistically evaluate these maritime giants, a range of metrics provides a comprehensive picture of their stature and impact:

Market Share

One of the most telling metrics, it offers a clear snapshot of their dominance in the maritime sector relative to competitors.

Environmental Initiatives

Given the current emphasis on sustainability, their endeavors towards reducing carbon footprints, adopting green fuels, and minimizing environmental degradation play a pivotal role in assessing their industry stature.

Operational Efficiency

Parameters like turnaround time, vessel utilization rates, and operational uptime provide insights into their operational prowess.

Customer & Stakeholder Satisfaction

Feedback from clients, partners, and other stakeholders paints a holistic picture of their market reputation.

Financial Health

Analyzing financial statements, profitability margins, and growth rates underscores their economic vitality.

Digital Transformation & Technological Adoption

Their inclination and capability to integrate cutting-edge technologies into operations provide a window into their future readiness.

The entities of Maersk, MSC, and CMA CGM Group not only serve as successful maritime organizations but also embody the highest standard of maritime achievement. Through the application of rigorous criteria and measurements, examining these entities allows us to not only comprehend their present level of prominence but also acquire vital insights into the factors contributing to their achievements. This accumulation of knowledge serves as a great resource for individuals aspiring to enter the marine industry or those with a keen interest in the subject matter.

Profiles of Maritime Trendsetters

Maritime is a dynamic and competitive field. Within this industry, certain companies have consistently managed to stay ahead of the curve, setting trends and charting the course for others to follow.

Maersk

One such monumental entity is Maersk, a company that has become almost synonymous with maritime excellence.

Background and Brief History

The A.P. Moller - Maersk Group, also known as Maersk, was established in 1904 by Arnold Peter Møller and his father, Captain Peter Maersk Møller. The company's inception took place in Svendborg, Denmark. Maersk was originally founded as a maritime enterprise. Since then, it has undergone significant diversification over the course of the last century and has transformed into a comprehensive conglomerate, including transportation, logistics, and supply chain operations. Presently, this pioneer company in the shipping sector demonstrates the visionary thinking of its creators and reflects the unwavering commitment to achieving high standards for its employees.

Innovative Strategies and Practices

Following are Maersk's innovative strategies and practices

Operational Scale

Maersk's growth can largely be attributed to its expansive global operations, linking all corners of the globe with its vast fleet and intermodal capabilities.

Innovation & Adaptability

Maersk has been at the forefront of digital transformation in maritime. With platforms like TradeLens, which leverages blockchain technology, it has revolutionized supply chain transparency and efficiency.

Economic Footprint

As indication of its economic prowess, Maersk has consistently recorded impressive revenues, making significant contributions to economies wherever it operates.

Strategic Alliances & Partnerships

Collaborations with tech giants, such as IBM for TradeLens, reflect Maersk's commitment to synergistic partnerships that amplify value creation.

Industry Leadership & Vision

Guided by its visionary leaders and driven by its commitment to sustainability, Maersk has pioneered initiatives like the goal to achieve carbon neutrality by 2050.

Metrics Underscoring Maersk's Influence and Success

Following are the factors that underscore Maersk's influence and success:

Market Share

Dominating several maritime service segments, Maersk's market share is indicative of its unmatched scale and reach in the industry.

Environmental Initiatives

Leading the green shipping revolution, Maersk's eco-friendly vessels and practices have set benchmarks for environmental responsibility.

Operational Efficiency

From state-of-the-art vessels to advanced port operations, Maersk's efficiency metrics often surpass industry averages.

Customer & Stakeholder Satisfaction

With a vast clientele that swears by its services and a reputation for reliability, Maersk's customer satisfaction levels remain enviably high.

Financial Health

Regularly posting robust financial figures, Maersk's economic health is a testament to its sound business strategies and operational excellence.

Digital Transformation

With digital tools like MyMaerskLine and Maersk Spot, the company has consistently demonstrated its capability to integrate technology seamlessly into its operations.

Key Outcomes and Impact on The Global Maritime Sector

Maersk's commitment to sustainable, efficient, and innovative maritime solutions has made it a beacon for other players. Whether adopting green fuels, pioneering digital platforms, or elevating global trade through efficient logistics, Maersk's impact on the maritime sector is profound and undeniable.

Maersk is more than a shipping corporation; it is a marine phenomenon. It guides the complex global marine sector with history, vision, and unrelenting innovation. By diving into Maersk, we see not only a maritime titan but also the blueprints for future nautical initiatives.

Mediterranean Shipping Company (MSC)

MSC emerges as a formidable entity whose impact should not be underestimated.

Background and Brief History

Mediterranean Shipping Company, commonly referred to as MSC, was established in 1970 by Gianluigi Aponte in Naples, Italy. Beginning with a single ship named Patricia, MSC has metamorphosed into the world's second-largest shipping line in terms of container vessel capacity. With its headquarters in Geneva, Switzerland, the company's growth trajectory epitomizes the vision realized through hard work, determination, and strategic planning.

Innovative MSC's Strategies and Practices

Following are Maersk's innovative strategies and practices:

Rapid Expansion and Agility

MSC's meteoric rise can be attributed to its capacity to rapidly expand its fleet, ensuring agility in operations and facilitating global coverage.

Commitment to Sustainability

Acknowledging the environmental imperatives, MSC has heavily invested in eco-friendly vessels and is known for its sustainable approach to shipping, often going beyond regulatory requirements.

Diversification

Beyond container shipping, MSC has diversified into cruise shipping, terminal investments, and logistics, indicating its ambition to touch every aspect of maritime services.

Customer-Centric Approach

MSC's services are renowned for their customer focus, with customizable solutions and dedicated support systems, ensuring client satisfaction and loyalty.

Metrics Underscoring MSC's Influence and Success The following are the factors that underscore MSC's influence and success:

Fleet Size & Capacity

MSC boasts one of the largest container fleets, reflecting its monumental scale and global reach.

Operational Excellence

Their port turnaround times and on-time delivery metrics often set industry benchmarks, a testament to their operational prowess.

Economic Contributions

With operations in numerous countries, MSC's economic contributions are substantial, driving growth and employment in multiple regions.

Stakeholder Engagement

A robust stakeholder engagement strategy ensures that MSC is not only responsive to feedback but also anticipates and caters to emerging demands.

Innovative Ventures

MSC's investment in digital tools and platforms exemplifies its forward-thinking approach and its commitment to revolutionizing the shipping experience.

Key Outcomes and Impact on the Global Maritime Sector MSC's indomitable spirit and its penchant for innovation have endowed the maritime industry with new benchmarks of excellence. Their green initiatives, extensive global network, and unwavering commitment to client satisfaction have inspired peers and have had a profound influence on maritime strategies worldwide.

Mediterranean Shipping Company stands as a colossus in the maritime sector, exemplifying how visionary leadership, coupled with innovative strategies, can propel a company to global prominence. MSC's story is not just that of a company's journey to the zenith of maritime trade but is also a lesson in innovation, resilience, and sustainable growth in a challenging global ecosystem.

CMA CGM Group

Amidst this cacophony of innovation and competition, certain names rise above, not just because of their scale but because of their significant contributions to the world of shipping. One such company, with its roots in France and global influence, is the CMA CGM Group.

Background and Brief History

Established in 1978 by Jacques Saadé in Marseille, France, the CMA CGM Group has grown from a single-ship operation to one of the world's leading maritime transport giants. The name "CMA CGM" amalgamates "Compagnie Maritime d'Affrètement" (CMA) and "Compagnie Générale Maritime" (CGM) – two major French shipping lines which merged in 1999. Within just a few decades, the Group expanded its footprint across continents, drawing strength from strategic acquisitions and a vision that prioritized both people and innovation.

Innovative Strategies and Practices

Following are CMA CGM Group's innovative strategies and practices:

Investment in Mega-ships

CMA CGM made headlines with its investment in ultra-large container vessels, recognizing the efficiency and environmental benefits of larger ships.

Focus on Sustainability

The group took significant strides in sustainable shipping, ordering a series of LNG-powered vessels, a first for such a scale in the industry.

Digital Transformation

With platforms like "CECIL," an online platform for e-commerce logistics, and "Shipfin," which offers financing solutions, the Group has embraced the digital age to enhance customer experience.

Diversification

Through acquisitions like CEVA Logistics, the company expanded beyond core shipping into comprehensive logistics solutions.

Metrics Underscoring CMA CGM Group's Influence and Success The following are the factors that underscore CMA CGM Group's influence and success:

Operational Spread

With a presence in 160 countries and a fleet of 500 vessels, CMA CGM's vast operational scale stands as a testament to its global prowess.

Commitment to Environment

Their pledge to not use Northern Sea routes to protect fragile ecosystems and the ordering of eco-friendly ships indicate their commitment to environmental responsibility.

Economic Contribution

With steady financial growth and a massive employment generation, the Group significantly impacts the global economy.

Customer-Centric Approaches

Consistently high ratings on customer service reflect their focus on delivering unparalleled service quality.

Innovation Index

Their continual investment in research and development, combined with partnerships with startups through incubators, signifies their commitment to remain at the forefront of innovation.

Key Outcomes and Impact on the Global Maritime Sector CMA CGM's emphasis on sustainability has influenced industry standards, prompting others to consider green alternatives more seriously. Their foray into digitization, diversified logistics, and customer-centric strategies have not only cemented their place as industry leaders but also paved the way for a more holistic view of maritime business beyond just shipping.

CMA CGM Group's journey is not just about maritime success but is a demonstration of visionary leadership, adaptability, and an unwavering commitment to innovation. By studying the Group's journey, maritime aspirants can glean insights into how traditional shipping can be blended with modern strategies to achieve unparalleled success in this ever-evolving sector.

Lessons Learned and Best Practices

In the vast expanse of the maritime sector, the waves of change are ceaseless, and the winds of competition are relentless. Crossing these waters requires not just great ships but visionary leadership, steadfast commitment, and innovative strategies. As we delve into the journeys of some maritime giants, there are a plethora of lessons to be gleaned and best practices to be emulated.

Overarching Themes from the Case Studies

The analysis of industry leaders such as Maersk, MSC, and CMA CGM Group unveils a number of themes that are both distinct and interconnected. Some of these include:

Visionary Leadership

Each company has been guided by leaders with a clear vision, leaders who weren't afraid to take risks and looked far into the horizon, anticipating shifts in the maritime sector.

Sustainability as a Priority

These companies recognize that environmental responsibility isn't just a necessity but a duty. Green shipping initiatives, sustainable fuel alternatives, and eco-friendly business practices have been central to their operations.

Embracing Digital Transformation

In an age of digital disruptions, these companies have strategically integrated technology into their operations, enhancing efficiency, transparency, and customer experience.

Success Factors and Common Strategies among Leading Companies

While the paths taken by these companies may vary, certain success factors emerge as common threads:

Continuous Innovation

Whether it's in the realm of propulsion systems, vessel design, or logistics management, a constant thirst for innovation has kept these companies at the industry's forefront.

Global Outreach with Local Insights

Expanding globally but understanding local dynamics, customs, and regulations have enabled these companies to establish a robust and widespread network.

Employee Well-being and Training

Recognizing that their greatest asset isn't just their fleet but their people, investments in training programs and employee welfare have been paramount.

Potential Pitfalls and How These Companies Faced Challenges

No journey is devoid of storms, and these maritime leaders have had their share:

Economic Fluctuations

Facing economic downturns, these companies adopted measures like cost-cutting without compromising on quality, diversifying revenue streams, and hedging against risks.

Stringent Regulations

With ever-evolving maritime regulations, these companies ensured robust compliance mechanisms, often exceeding mandatory requirements.

Supply Chain Disruptions

Whether due to geopolitical reasons or global crises like pandemics, agile supply chain strategies, contingency planning, and leveraging technology have been vital in ensuring business continuity.

The maritime account of giants like Maersk, MSC, and CMA CGM Group offers more than just business insights—it offers a compass for navigation in turbulent waters. By understanding their journeys, their strategies, their challenges, and their victories, maritime entities, whether old or new, can chart a course toward a brighter, more sustainable, and more prosperous future. The essence lies in learning, adapting, and constantly evolving.

The Role of Innovation, Sustainability, and Collaboration in Their Success

The triad of innovation, sustainability, and collaboration has been instrumental:

Innovation

Beyond technology, innovation for these companies spans processes, management practices, and even foreseeing market demands.

Sustainability

No longer just a buzzword, sustainability has become an integral business strategy, with benefits ranging from operational cost savings to enhanced brand reputation.

Collaboration

Realizing that the maritime challenges are vast and complex, these leaders have often collaborated—be it with tech startups, research institutions, or even erstwhile competitors—to achieve common goals.

Predictions from These Trendsetting Companies' Directions

Drawing from their past and present, we can predict:

Greater Integration of Tech

From AI-driven logistics management to IoT-enabled ship maintenance, technology will play an even more pivotal role.

Enhanced Focus on Green Shipping

With a global emphasis on sustainability, eco-friendly maritime operations will not just be preferred but essential.

Diversification of Services

Beyond mere shipping, these leaders might venture deeper into sectors like offshore renewable energy, tapping into the blue economy's potential.

From our analysis, it becomes evident that the future maritime sector will be one that's more interconnected, technology-driven, and environmentally conscious. These trendsetters haven't just shown us what's possible but have also set benchmarks for what should be the new normal.

Charting a Course Inspired by Leaders

In the marine domain, prominent entities such as Maersk, MSC, and CMA CGM Group emerge as beacons, shedding light on avenues of ingenuity, steadfastness, and exceptional performance. The experiences of these individuals highlight the profound impact that visionary leadership and perseverance can have. For emerging maritime entities, these narratives serve as both a source of motivation and a strategic guide, encouraging them to adjust, innovate, and envision a future in which the maritime domain encompasses not only the navigation of oceans but also the establishment of sustainable and inventive methodologies, thus contributing to a more promising outlook for the entire industry.

Chapter 9

CONCLUSION AND
FUTURE VISIONS

Synthesizing the Maritime Journey

As we approach the last stages of our enlightening maritime expedition, let us pause to anchor ourselves and contemplate the progress we have made thus far. The maritime industry, which serves as a fundamental pillar of global trade and interconnectivity, is currently experiencing a period of significant transformation. Our exploration of the chapters has revealed not only the industry's complexities but also its adaptation to the ever-changing global dynamics.

Recapturing Key Insights from Previous Chapters

Our journey began with understanding the foundational structures of the maritime industry, its economic significance, and its challenges. From there, we delved into the paradigm shifts propelling the industry towards sustainable operations. We explored global trends highlighting the importance of sustainable shipping and the immediate and long-term benefits of green shipping. Our path took us through the alternative fuels, propulsion systems, and game-changing initiatives of maritime giants like Maersk, MSC, and CMA CGM Group. These companies, with their innovative approaches, set the stage for understanding the importance of adopting sustainable and technologically advanced practices.

As we continued our exploration, the role of economic forecasts and business models became apparent. The maritime industry, in its anticipation of the future, is clearly seen to be evolving its business models, integrating technologies like blockchain, and emphasizing the circular economy. We also learned about the significance of collaborative efforts and alliances in propelling green shipping initiatives.

Reflecting on the Global Shifts in Maritime: Sustainability and Technological Advancement

The narrative of our exploration unmistakably underscores two key global shifts in the maritime sector: Sustainability and Technological Advancement.

The concept of sustainability has transitioned from being a mere buzzword to becoming an imperative requirement. The increasing recognition of the environmental consequences associated with conventional maritime activities has led to a heightened global focus on adopting environmentally sustainable methods. This not only addresses the environmental concerns but also makes economic sense, as we've learned from the benefits green shipping brings in terms of cost savings and future-proofing against stricter regulations.

On the technological front, the maritime industry is not untouched by the fourth industrial revolution. The adoption of digital technologies, automation, and the integration of AI and machine learning into operations are more than just trends; they are reshaping the entire maritime landscape. From streamlining operations and improving efficiency to ensuring safety and enhancing customer experiences, technology is at the helm of the Maritime's journey into the future.

Together, these global shifts are not just reactive measures. They represent the maritime industry's proactive approach to a future that is sustainable, efficient, and technologically advanced. As we synthesize our journey, it becomes clear that the maritime industry is at an inflection point, poised to set sail into a future that promises innovation, sustainability, and immense growth.

Embracing Change as the New Norm

Within the dynamic framework of the marine domain, continuous transformation remains the sole lasting feature. The oceans, characterized by their tides, currents, and inherent unpredictability, provide as a tangible representation of the ever-changing nature of this field. Throughout the course of human history, it becomes evident that individuals who are able to adeptly navigate the constantly shifting environment are the ones who achieve success in this vast domain.

Understanding the Imperative for Adaptability in Maritime

The maritime industry, with its deep-rooted traditions and century-old practices, might seem static to the untrained eye. However, as global commerce expands and environmental and technological pressures mount, adaptability isn't just a virtue; it's a dire necessity. The acceleration of digital transformation, stringent environmental regulations, and changing global trade dynamics underscore the need for adaptability.

For maritime entities, adaptability isn't about abandoning age-old traditions but about integrating them with new-age practices. It's about combining the wisdom of the past with the innovations of the present to chart a course for a prosperous future. Whether it's the shift to green shipping, the adoption of digital tools, or the strategic alignment with global trade shifts, adaptability is the anchor that ensures stability amidst the changing tides.

Success Stories of Companies That Pioneered Change

Maersk

A notable exemplar of embracing change, Maersk's journey from a conventional shipping company to a digital leader in logistics and services demonstrates adaptability in action. Their commitment to reducing CO2 emissions and their investment in digital solutions, such as TradeLens, showcases their foresight in pioneering industry change.

CMA CGM Group

With its acquisition of CEVA Logistics, CMA CGM Group exemplified how integrating logistics solutions can expand horizons. Their dedication to LNG-powered vessels and their forward-thinking approach to eco-friendly shipping solutions paints a portrait of a company willing to redefine industry norms.

MSC

MSC's green initiatives, especially their commitment to fully carbon-neutral operations, have set them apart. Their belief in continuous evolution, showcased by their investments in alternative fuels and advanced waste treatment systems, reinforces the narrative of a company that sees change not as a challenge but as an opportunity.

Each of these case studies offers a narrative of triumph, not just in terms of profitability, but in pioneering change. These companies didn't just adapt; they became the vanguards of transformation, showcasing to the world that in the face of change, the maritime industry is not just resilient but revolutionary.

The Crucial Role of Innovation and R&D

The maritime domain, characterized by its expansive nature and enigmatic qualities, is an analogy for the unexplored regions of innovation. As we look toward the horizon of the future, it's evident that the ships powered by the dual engines of innovation and R&D will sail smoother, faster, and with greater purpose. Both have become the lighthouses guiding the maritime sector, ensuring it not only stays relevant but also thrives amidst the evolving challenges and opportunities.

Technology as the Future of Maritime

The role of technology in reshaping the maritime industry cannot be understated. Digital transformation, encompassing everything from AI-driven logistics to IoT-enabled ships, is revolutionizing operations, safety, and efficiency. Satellite communication is ensuring seamless connectivity, even in the heart of the vast oceans, while autonomous vessels hint at a future where human error on seas could be minimized.

Beyond just operational efficiency, technology promises sustainability. Advanced hull designs, energy-efficient engines, and emission control technologies are making ships more environmentally friendly. With the looming deadlines of international environmental regulations, technological innovations aren't just luxuries; they are necessities.

Why Investment in R&D is Non-negotiable

In a rapidly evolving global landscape, companies that stand still risk becoming obsolete. The purpose of Research and Development (R&D) extends beyond mere competitive advantage since it involves the fundamental reevaluation and transformation of established norms and regulations within a given field. For the maritime industry, the stakes are even higher. Environmental concerns, safety protocols, and the need for cost efficiency demand constant innovation.

The distribution of financial resources is not the sole consideration when it comes to investing in research and development (R&D). The focus lies on cultivating a culture that encourages curiosity, unwavering dedication to improvement, and a steadfast commitment to future initiatives. R&D ensures that companies don't just react to changes but anticipate and shape them.

Anticipating Market Needs through Proactive Research and Development

True industry leaders don't merely respond to market demands; they anticipate and create them. Through proactive R&D, maritime companies can gauge emerging trends, understand potential disruptions, and develop solutions even before challenges manifest. It's about having a pulse on everything, from evolving geopolitical scenarios that might impact trade routes to breakthroughs in material science that could redefine shipbuilding.

Furthermore, by actively engaging in research, companies can discover untapped opportunities, whether they lie in new maritime routes, unexplored logistical solutions, or innovative service offerings. In essence, proactive R&D transforms maritime entities from being players in the game to being the very architects of the industry's future.

The Rise of the Learned Maritime Professional

As the tides of the maritime sector shift, they bring with them a strong current of change, not just in technology and practices but also in the caliber of professionals steering the industry. Today, the spotlight shines brightly on the learned maritime professional—a beacon of knowledge, adaptability, and foresight. This new breed of maritime expert understands the waves of change and possesses the skills to surf them proficiently.

What Does Upcoming Maritime Leaders Need

Tomorrow's maritime leaders will need a toolbox of skills, some traditional and others newly emergent. Of course, foundational knowledge about shipping, logistics, and marine ecosystems remains paramount. But as the sector gets increasingly intertwined with technology and global socio-economic dynamics, a broader skillset becomes essential.

Digital proficiency, for instance, is no longer optional. Understanding AI-driven logistics, digital supply chain management, and cybersecurity will be as foundational as knowing the ins and outs of a ship. Furthermore, with sustainability at the helm, leaders will need expertise in green technologies, environmental regulations, and sustainable business practices.

In addition to technical expertise, the acquisition of soft skills such as cross-cultural communication, leadership in varied environments, and strategic vision will be of utmost importance. In light of the increasing interconnectedness of the marine industry, its leaders must possess the requisite skills to effectively negotiate the complicated issues associated with global collaborations and partnerships.

Strategies for Continuous Professional Development in the Sector

Continuous learning is the anchor that will ground maritime professionals in this sea of change. Here are some strategies to ensure ongoing professional growth:

Lifelong Learning Platforms

Many institutions and organizations now offer specialized courses tailored to the maritime industry, covering everything from technological advancements to sustainability practices. Leveraging these platforms will be crucial.

Industry Seminars and Workshops

Regular participation in workshops, webinars, and seminars will keep professionals updated with the latest trends, challenges, and solutions.

Networking

Engaging with industry peers, joining maritime associations, and attending conferences can provide insights beyond formal education. The exchange of on-ground experiences and best practices is invaluable.

On-the-Job Training

Companies can play a pivotal role by investing in training programs, ensuring their workforce is equipped with the latest skills and knowledge.

Cross-industry Exposure

Given the interdisciplinary nature of modern maritime challenges, exposure to other sectors like technology, environment, and finance can provide a broader perspective and innovative solutions.

In wrapping up this section, it's evident that the maritime sector, with its intricate blend of tradition and innovation, calls for professionals who are both rooted and visionary. The learned maritime professional, with a compass of knowledge and a map of continuous learning, is set to lead the industry into its golden age.

Call to Action: The Path Forward for the Maritime Industry

With the significant changes in technology, environment, and economy, the maritime industry stands at a critical point. The decisions made in the present time will help the industry to withstand challenges and achieve success in the forthcoming decades. They will also influence worldwide trade, economies, and ecosystems. This book functions as an appeal for unity to all individuals and companies involved, setting forth a clear goal and outlining a path for progress.

Cultivating a Culture of Innovation and Learning

The maritime industry's evolution is a deeply ingrained culture of innovation and continuous learning. As maritime businesses confront unprecedented challenges, a proactive and forward-looking mindset becomes a linchpin for success. Key facets of this culture include:

Encouraging Experimentation

Companies must create an environment where experimentation is rewarded and failures are seen as learning opportunities. This ethos fosters the development of groundbreaking solutions.

Dedicated Innovation Labs

Investment in dedicated research and development labs can help businesses remain at the cutting edge of maritime technology and practices.

Continuous Learning Platforms

By integrating continuous education into their organizational DNA, maritime companies can ensure their workforce stays ahead of industry shifts and is well-equipped to adapt to change.

A Vision for a Sustainable, Technologically Advanced Maritime Future

The maritime future we envision is one where sustainability and technological advancement sail hand in hand. It's a world where:

- Green shipping practices are not mere compliance strategies but form the core of all maritime activities.
- Digitalization, from AI-driven logistics to blockchain-based supply chain management, redefines efficiency and transparency in the sector.
- The maritime carbon footprint diminishes significantly, thanks to a massive adoption of alternative fuels and propulsion systems.

The Role of Collaborations, Alliances, and Partnerships in Shaping the Future

No single entity can navigate the maritime future alone. The complexity and scale of challenges necessitate collective action. Collaborations, alliances, and partnerships will play a central role in:

Pooling Resources

Joint ventures can allow stakeholders to share risks, costs, and expertise—accelerating the development and adoption of innovative solutions.

Standardizing Practices

Through alliances, industry leaders can set benchmarks and standards, ensuring uniformity in best practices across the globe.

Knowledge Exchange

Collaborative platforms can facilitate the sharing of research, insights, and experiences, driving the collective growth of the sector.

The maritime sector's journey ahead, though laden with challenges, is rife with opportunities. By embracing innovation, prioritizing sustainability, and championing collaboration, the industry is not just poised to navigate the turbulent waters but to redefine its very horizons. This is not just a vision but a call to action, urging every stakeholder to take the helm and steer towards this promising future.

Final Reflections: The Odyssey of the Maritime Industry

This odyssey, a blend of past practices and future aspirations, encapsulates the essence of an industry in flux yet firmly anchored in its core values.

Tradition and Modernity in Maritime

The maritime industry, dating back millennia, is one of the world's oldest. It carries the weight of time-honored traditions, customs, and practices. However, as with any ancient industry, there's a dance between holding onto these traditions and embracing modern innovations:

Respecting Heritage

The foundational practices and values of maritime, forged through centuries, offer a compass that guides even in turbulent waters. This respect for tradition grounds the industry and offers a unique identity.

Embracing Innovation

While tradition anchors, innovation propels. The maritime industry, in its quest to remain relevant, has harnessed technological advancements, from satellite navigation systems to autonomous vessels, making the age-old practices more efficient and environmentally friendly.

Emphasizing the Collective Responsibility of Professionals and Companies

The maritime industry doesn't operate in isolation—it's a collective endeavor. The choices made by professionals and companies alike have ripple effects, influencing global trade, marine ecosystems, and the broader environment:

Professionals as Custodians

Every individual in the maritime sector, from the ship captain to the logistic manager, holds a piece of the industry's legacy. Their decisions, ethics, and actions shape the industry's present and future.

Corporate Accountability

Companies, as the titans of the maritime world, bear a significant responsibility. Their strategies around sustainability, employee welfare, and technological adoption set the tone for the industry at large.

Embracing Challenges as Opportunities

Every era presents its own set of challenges. For the maritime industry, these range from environmental concerns to technological disruptions. However, the industry's resilience lies in its ability to view challenges as catalysts for growth:

Adaptive Mindset

The maritime sector's history is replete with examples of adaptability—be it the shift from sails to steam or adapting to new trade routes. This inherent adaptability ensures the industry's survival and prosperity.

Opportunity in Adversity

Challenges, be it stringent regulations or economic downturns, often spur innovation. They force the industry to introspect, innovate, and emerge stronger.

In conclusion, the maritime industry's odyssey reflects its enduring spirit. As it charts the waters of tradition and modernity, faces challenges, and evolves, one thing remains constant—its unwavering commitment to growth, progress, and global connectivity. As we set our sights on the future, we do so with a profound appreciation for this remarkable journey and an optimistic anticipation for the horizons yet to be explored.

www.ingramcontent.com/pod-product-compliance
Lightning Source LLC
Chambersburg PA
CBHW080851120626
46546CB00008B/2779